# WORDS FROM THE WILDWOOD

## Cumbria's Ancient Place-names

# WORDS FROM THE WILDWOOD

## Cumbria's Ancient Place-names

# Robert Gambles

*with photographs by Val Corbett*

HAYLOFT PUBLISHING LTD
CUMBRIA

This new and revised edition published by Hayloft Publishing Ltd., 2017
First published as *Out of the Forest*, Laverock Books, 1989

A CIP catalogue record for this book is available from
the British Library

ISBN 978–1–910237–30–4

Designed, printed and bound in the EU

Hayloft policy is to use papers that are natural, renewable and  recyclable
products and made from wood grown in sustainable forests.  The logging and
manufacturing processes are expected to conform to the environmental
regulations of the country of origin.

Hayloft Publishing Ltd,
a company registered in England number 4802586
2 Staveley Mill Yard, Staveley, Kendal, LA8 9LR (registered office)
L'Ancien Presbytère, 21460 Corsaint, France (editorial office)

Email: books@hayloft.eu
Tel: 07971 352473
www.hayloft.eu

*In memory of my wife, Hannemor,*
*whose love made everything possible*

O! who can tell
The hidden powre of herbes and might of Magick spell?

Edmund Spenser, *The Faerie Queen*

# CONTENTS

## PREFACE TO THIS EDITION

MUCH has changed in the thirty years since I wrote *Out of the Forest*, the first edition of this study of Man's relationship with the Natural World as reflected in the place-names of Cumbria. A number of birds and animals then regarded as extinct in England have since returned or been successfully reintroduced: birds such as the red kite, the osprey and the crane; animals such as the beaver, the wild boar and the pine marten. Others such as the skylark and golden plover, the hedgehog and the red squirrel, are now on the list of endangered species.

Similarly several once common flowering plants and herbs of the fields, hedgerows and woodlands have suffered from modern farming practices and commercially motivated coniferous forestry while trees such as the elm, ash and oak are succumbing to attack from devastating diseases. Insofar as these changes have made an impact on Cumbria appropriate amendments have been made in the text of this new edition but perhaps more important is the additional emphasis given to the underlying theme of the book, namely the dependency of our medieval ancestors on their knowledge of and intimate relationship with the natural world around them.

The opportunity has also been taken to present the book with a new title.

It is with pleasure that I express my appreciation of the help which I have received in the preparation of this edition from the librarians in Kendal and Carlisle, from the Cumbria Wildlife Trust, from Dr Frank Steven, and from my son, Peter, whose computer skills came to the rescue on many occasions.

I wish also to thank Val Corbett for providing the cover illustration and Dawn Robertson, my publisher, for her enthusiastic encouragement and guidance.

Robert Gambles, 2017

# PREFACE TO THE ORIGINAL EDITION

IN the introduction to his *English Social History* G. M. Trevelyan defined his subject as the daily life of the inhabitants of the land in past ages and he included among the essential elements of this the attitude of man to nature. Today we have become much more aware of the importance of our environment and, in particular, of our relationship to the natural world about us. To this extent we have regained a little of that careful perception of Nature which distinguished earlier generations whose daily lives were so dependent on the resources which Nature had provided, who often lived in awe of its elemental forces and who were never in doubt that the natural world had a life of its own in no way dependent on the needs of humanity. They had no choice but to live in conditions of close intimacy with the physical features of their landscape and with the vegetation and wildlife in the boundless forests surrounding their tiny homesteads. To attempt to recreate a glimpse of this lost world is part of the historian's task and it is as an historian that I have presented these reflections on some of the place-names of Cumbria.

Place-names studies are a minefield for the unwary, frequently fraught with uncertainty, so tantalisingly complex and incomplete is the linguistic and documentary evidence. Equally uncertain is the problem of accurately identifying some of the plants referred to in ancient sources. From Greek and Roman times until the sixteenth century botanical studies were based on the perceived needs of medicine rather than on any botanical science, and as late as 1568 William Turner's *Herbal* placed emphasis on 'the Doctrine of Signatures', the belief that God 'hath imprinted on the plants ... the very signatures of their virtues'. Thus the spots on the leaves of pulmonaria (lungwort) were taken as a sign that this plant was to be used for the treatment of diseased lungs, and the nodules on the roots of the lesser celandine indicated its value in the treatment of piles.

This Tudor admixture of charlatanry to herbal medicine does not help our problem, interesting though it is, and we are fortunate to have more reliable sources to give us contemporary insight into the beliefs and practices of the peoples who colonised much of Britain over a thousand years ago and created almost all the place-names referred to in this book. Thus a twelfth century *Danish Herbal* by Hendrik Harpestreng and several Anglo-Saxon Leechdoms leave us in no doubt that curiosity about the natural world at that time was directly related to the practical problems of daily life.

The Victorian writer, R. C. A. Prior in his book *On the Popular Names of British Plants*, noted that our fruit and timber trees, the cereal grains, and several potherbs and medicinal plants, have the same names at the present day as they bore a thousand years ago and observed that earlier generations gave special names only to those objects of Natural History which were conspicuously useful, beautiful or troublesome. It is not surprising, therefore, to discover that this is exactly reflected in the place-names of the Celtic, Norse and English settlements. There were certain plants, trees, birds and animals whose habitats it was important, often a matter of survival, to be able to identify.

It is worthy of note that less than 200 years ago a Doctor

Batter, medical practitioner of Market Lavington in Wiltshire, treated his patients by prescribing not only the names of the plants or herbs which would provide the necessary medicine but also the precise location where they could be found; and in Elizabeth Gaskell's novel *Mary Barton,* published in 1848, the Cumbrian-born Alice Wilson went out all day in the fields gathering wild herbs for drinks and medicines.

I am grateful to all those who have offered help and guidance in the writing of this book and I should like especially to record my thanks to the librarians in Kendal, Carlisle and the University of Lancaster who gave so generously and patiently of their time and expertise in the search for references and illustrations.

Robert Gambles, 1988

# NOTE ON SHAKESPEARE

QUOTATIONS from Shakespeare's plays enliven many of the chapters in this book. Shakespeare lived at a time when many of the beliefs, superstitions and customs of the early and medieval centuries were still generally prevalent. He himself was very familiar with the folklore of the countryside and it seemed appropriate to accept his many references to it as illustrations of contemporary and traditional perceptions of the natural world.

# INTRODUCTION

THE place-names of Cumbria were created by the men and women who first established their homesteads in the forests which covered almost all the region until the sixteenth century. From their first tiny clearings they explored the countryside around them and got to know in intimate detail every fell, crag, moss, beck and woodland, and all the animals, birds, trees and plants which lived there. Whether they were Celts, Anglo-Saxons, Norsemen or medieval Englishmen, they all had one primary purpose in their daily lives: to survive in a harsh and largely hostile world. Life was, for many, an unrelenting struggle for survival against hunger, cold, disease and an early death, and it was this, above all, which dominated their day-to-day existence and determined their relationship with and attitudes to the natural world.

Nature's bounty provided them with food, fuel, and the raw materials with which to build their homes and make their clothing, and also with the 'magic' herbs and plants which could ease their many physical ills. But Nature's caprice could, equally, wreak sudden havoc and destroy an entire harvest; carefully nurtured crops could be ravaged overnight by ravenous birds and wild animals. Storm and flood could, in a few short hours, sweep away food, flocks, mills, byres and all they possessed. Sheep, cattle, goats, geese and hens so easily fell prey to the beasts of the forest, thus depleting their sources of meat, milk, cheese and eggs. Hunger, disease and the hazards of childbirth carried off young and old with frightening frequency. Throughout these medieval times country folk lived in a proximity to Nature so close that we can scarcely imagine it: Nature allowed them to live but its power was to be feared and respected and its charity garnered with the knowledge and skills passed on from generation to generation.

So these early farmer-settlers took careful note of those places where Nature either helped or hindered them in the struggle to survive. They discovered where the trees grew which could best be used for fuel, for building or for the manufacture of farm and household implements; they sought out and gave names to the places where they could find edible plants and medicinal herbs; they found out the dens and haunts of wolf, fox, wildcat and boar which killed their sheep and poultry and ravaged their crops; they named the crags where eagle and raven kept watch for sickly calves and new-born lambs; they learned the nesting places of blackcock, heron and dove and all the birds they could snare for the pot; they identified the waters where salmon, trout and eel abounded. They noted everything in the countryside that was friend or foe, edible or inedible, useful or useless; and, as any dialect dictionary or place-name study will show, their perception and observation of detail was remarkable and encyclopaedic.

Place-names can help to lift the veil on this world of 1,000 and 2,000 years ago when our ancestors lived in such intimacy with Nature. We may not be able to reconstruct a complete ecology from these names but we can, even within the uncertainties of place-name studies, rediscover something of a way of life which for many ordinary folk survived in part until the eighteenth century or, for some, even later. Almost all these names appear in their first written form in the medieval centuries before the beginning of the transformation in Man's relationship with Nature which was to result not merely in Man's conquest of Nature but in his ruthless and overbearing exploitation of it.

Population pressures created a demand for greatly increased supplies of food and soon only cultivated land had any virtue. Acts of Parliament in the sixteenth century commanded every parish to destroy as 'vermin' a vast catalogue of animals and birds. Flowers and plants not cultivated in field or garden were despised as 'weeds'; woodlands were regarded mainly as a source of fine timber for the ships of the new Royal Navy or as a plentiful supply of fuel for the maw of the rapidly expanding iron and textile industries. The agricultural revolution speeded up the transformation while the invention of the sporting gun made possible the mass slaughter of every form of wildlife.

The growth of industrial towns tore up by its roots much of the ancient culture of the countryside and, with it, much of that fund of inherited knowledge which, a few short years before, had sent the new scientific botanists and biologists in astounded wonder to humble cottages and farms in search of information on the species, habitats and properties of so many living things. They discovered not only much that could be catalogued, dissected and classified, but also much accumulated folk-wisdom that science could not then explain: the secrets of country medicines which had survived since the old Anglo-Saxon Herbals and even before; the power of the rowan and the hawthorn to hold the forces of evil at bay; the mystery surrounding the sinister behaviour of the cuckoo and the toad; the magic healing powers of the ash-tree and the supernatural virtues of the holly and the oak.

This and much more was part of a pagan folklore common

to almost all the peoples of northern Europe, often adopted and modified by the Christian Church, and certainly an inherent part of the customs and beliefs of the pioneer settlers who gave names to the hundreds of places in Cumbria which refer to the teeming natural world within which they lived. Nineteenth century science and urban society did much to end all this but the Cumbrians of Victorian England still placed rowan twigs over their doors against witchcraft and even today the belief is still widely held that bad luck will surely follow if an elder tree is cut or its wood burned.

# NOTE ON THE PLACE-NAMES

THE place-names given in each chapter are not presented as a comprehensive list but rather as an illustrative selection. Most of the names are readily found on the standard (1:25,000) Ordnance Survey Maps of Cumbria but, occasionally, field-names and names now lost have been included. In some instances the list would have become a tedious catalogue involving the repetition of many names with the same root construction. Where a name occurs which has a version recorded at an unusually early date this is indicated by the date in brackets after the name. These names can often be the most valuable sources of information for the correct explanation of a place-name's origin.

The study of place-names is a specialised academic discipline demanding a knowledge of several ancient languages and considerable expertise in the finer points of linguistic development. It is a field of studies into which the amateur ventures at his peril and the author fully acknowledges his debt to the many authoritative works which have been published in the past fifty years. Outstanding among these are the publications of the English Place-name Society and in particular the volumes dealing with the place-names of Cumberland and Westmorland and, more recently, Professor Diana Whaley's *Dictionary of Lake District Place-Names*.

Similarly the number of books and specialist works touching on the various aspects of natural history and folklore is immense and there is also a steadily increasing volume of work on the history of medicine and on the history of the Anglo-Saxon and Viking ages both in this country and in Scandinavia. Mastery of any one of these fields of study comes only to the life-long enthusiast and many of the themes of this book have been the subjects of detailed and scholarly treatises and keen amateur observation.

It is often the function of the historian to amass a sum of information on many aspects of the past and, with the intellectual discipline of his profession to try to reconstruct from innumerable sources the daily life of our ancestors. One of these sources is the vast number of place-names given to every corner of the land by the early generations of settlers, a difficult but unique fund of detailed information which Sir Walter Scott may well have regarded as a characteristic preserve of his historical researcher *Dryasdust*. One need not share his opinion that the smallest fact about the past was the true poetry to appreciate that even the linguistic revelations of place-name study can add much to our knowledge of a vanished age, and not the least significant part of this is the aspect of social history which Trevelyan called the attitude of man to nature.

# NOTE ON THE ILLUSTRATIONS

IT seemed appropriate to illustrate a book of this kind with many of the woodcuts which first appeared in the sixteenth century herbals and in the studies of birds and animals which were produced in the following two hundred years. Executed in the infancy of botanical and biological science these drawings may not always meet the exacting demands of modern studies or appeal to eyes more accustomed to brightly coloured reproductions. Their merit lies in that they were drawn directly from nature as 'living drawings' and were not intended to be a means of identification – this was the function of the accompanying text.

Of the English herbals the best known is *The Herball or General Historie of Plantes* by John Gerard first printed in 1597, a massive work of over 1,400 large pages and more than 2,000 woodcuts. Some difficulties arise in the precise identification of a few of Gerard's drawings owing to modern changes in classification and scientific nomenclature but this is, in truth, of little real significance as Gerard, like most folk at that time, was more concerned with the 'virtues' of a plant than with its botanical detail. Gerard's book is most usually seen in an edition published by Thomas Johnson in 1633 in which some of Gerard's artistic errors were put right, although Johnson's 'amendments' were not nearly as extensive as he claimed. He reprinted Gerard's text precisely adding only a few footnotes. This is the edition used here.

A few of the illustrations are taken from works contemporary with those of Gerard and Johnson:

> Leonhard Fuchs – *De Historia Stirpium* (1542)
> Petrus Andreas Matthiolus – *Commentarii...* (1565)
> Theodorus Zwinger – *Theatrum Botanicum* (1744)
> with woodcuts from Conrad Gesner made in the sixteenth-
> century.

The illustration of the aurochs is a copy made in the early nineteenth century by Charles Hamilton-Smith of what is believed to be a sixteenth century painting known as the Augsburg Aurochs.

The bird illustrations are from Thomas Bewick's *History of British Birds* (1804); the animal illustrations are from Bewick's *General History of Quadrupeds* (1807) and from Edward Topsell's *Historie of Foure Footed Beastes* (1607); the fish are from Isaac Walton's *Compleat Angler* (1653) and from John Hill's *An History of Animals* (1752); and the insects are from *The Theater of Insects* by Thomas Moffet (1658). The illustrations for the section on domestic animals are reproduced from the book edited by B. Cirker entitled 1,800 woodcuts by Thomas Bewick and his school (1962).

# 1
## FLOWERS AND HERBS

# ANGELICA

*Gerard*

**Derivation:** Old Norse – *hvønn*
**Place-name:** Wanthwaite, Wannethwayt (1301)

Known today chiefly for its candied stems used in confectionery and for its seeds used in the preparation of vermouth and chartreuse, angelica was once highly regarded as the source of herbal medicines which, it was believed, would be a remedy for many ailments. It was also believed to have the power to break spells and curses and to be a defence against

2

evil spirits. Angelica is one of the few aromatic plants to thrive in the Arctic regions and it was particularly highly esteemed among the Norsemen. Even today one of its alternative names is Norwegian Angelica.

In Iceland and Lapland the stems and roots were eaten raw and in Finland they were baked first, while in Norway they were also used in bread-making. Angelica has always grown in the wild in most countries of Northern Europe but among the earliest records of the Laws of Norway we find provision made for its actual cultivation both as a vegetable and for its medicinal benefits. This has survived in certain Norwegian place-names. The example given from Cumbria also indicates cultivation: Wanthwaite may be interpreted as 'the thwaite or cleared enclosure where angelica was grown'.

As a medicine the juice of the angelica root and a solution made from the seeds were commonly used as an eye lotion, and the powdered root in a solution made from the juices of the leaves and stalks was believed to bring relief for bronchial and rheumatic troubles. Above all, angelica was one of the many herbs used to treat digestive problems, a very common complaint in the days of primitive hygiene, lack of sanitation and an unbalanced diet.

Excavation of the waste pits at the Viking settlement at Jorvik in York has revealed that the folk who lived there suffered from gut-worm infection, with consequent stomach disorders for which herbs such as angelica provided a remedy. John Gerard, writing in 1597, and Nicholas Culpeper, fifty years later, confirm that in their day, too, its medicinal virtues were highly valued, adding for good measure, that 'it cureth the biting of mad dogs and other venomous beasts', a claim as improbable as Culpeper's statement that it was effective

3

against the plague and all epidemical diseases.

Improbable or not, veneration for angelica had an ancient history. It had long ago been incorporated into the pagan festival of Spring and was deeply rooted in folklore, so much so that the Christian Church adopted it for the Feast of Michael the Archangel (8 May) and added the name Archangelica, a name now used by botanists to distinguish *Angelica Archangelica* from the similar but less potent plant of our woodlands and hedgerows, *Angelica Sylvestris*, whose purple stems and blushing white flowers adorn the summer months.

# BARBERRY

*Zwinger*

**Derivation:**   Old French/Latin: *berberis*
**Place-names:**   Barbara Crag, Barbara Field, Barbara Rigg

Writing in the mid-seventeenth century Nicholas Culpeper commented:

> This shrub is so well known to every boy and girl that hath but attained to the age of seven years, that it needs no description.

The barberry must have spread very quickly to achieve such fame for in the Middle Ages it seems to have been known only in the herb gardens of monastic foundations whence it occasionally escaped to nearby hedgerows and private gardens. It soon became unpopular with farmers as it was identified as the villain causing wheat mildew spread by the spores of a fungus which attached itself to the barberry bush. It is also mildly poisonous but its more helpful virtues included its use as a medicine against fevers and as an anti-septic and for purging the body of choleric humours – the latter involving the drinking of a quarter of a pint of white wine in which barberry bark had been boiled, not, therefore, likely to be in everyman's medicine chest.

A mixture made from the lye of the ashes of the barberry root was used to dye linen and leather and also as a hair-shampoo which allegedly turned the hair yellow to achieve the fashionable blonde appearance which so many ladies craved. In the kitchen the barberry leaf featured as a seasoning to meat and as a salad but the berries which are exceptionally rich in Vitamin C were especially sought out for that favourite medieval conserve, barberry jelly, recommended by Mrs. Beeton in Victorian times as a dry sweetmeat and in sugarplums, but famous in gourmet circles as the confiture *d'épine-vinette* made only in Rouen.

The folk of medieval Cumbria probably knew nothing of such delicacies for the barberry was a rare plant in northern counties until modern times and it seems unlikely that it was ever common except as an ornamental garden shrub. Only three place-names are recorded and all are of a very late date and, surprisingly, all are grouped together adjacent to the Old Coach Road across the northern reaches of Matterdale Common.

# BILBERRY

*Zwinger*

**Derivation:**    Old Norse: *blá-ber* – dark blue berry
**Place-names:** Blaburthwaite (1272), Bleaberry Knott, Bleaberry Fell, Bleaberry How, Bleaberry Tarn, Bleaberry Gill, Bleaberry Haws

The bilberry once grew abundantly in the woodlands and on the lower fell-sides in Cumbria but in recent decades intensive grazing by sheep has meant that the young shoots are eaten off and so berries rarely have the opportunity to form. The introduction of managed Forest Commission enclosures now offers hope of a modest recovery.

7

In spring-time the delicate, drooping pink flowers of the bilberry provide a rich harvest for the honey bees; in the autumn its luscious dark blue berries are enjoyed by both humans, birds and some small mammals. It is mainly a fruit of the north and west, little known in the south and east. John Gerard commented from his Holborn garden that bilberries were 'eaten in the north with cream or milk as in these south parts we eate strawberries'.

No-one who has acquired bilberry-stained hands (or tongues) after picking bilberries will doubt that the juice was once used as a purple dye for linen or wool, but there can be few now living who can recall the days when bilberry leaves were used to make a tea allegedly invaluable as a nerve tonic and soothing for stomach upsets. Traditionally bilberries were used to treat mouth ulcers, kidney problems and weak eyesight; in modern medicine bilberry is recommended for the treatment of age-related macular degeneration of the retina

Bilberries are now hard to find in England but in Iceland and Scandinavia they are still abundant and are gathered in the forests each autumn to be eaten with cream or to be used in crêpes or made into pies, jam and liqueur. Bilberries with cream are no longer a common Lakeland summer delight except perhaps for the adventurous few who seek out the places where sheep prefer not to go. To our medieval ancestors they were a welcome seasonal fruit – with or without cream.

## BLACKBERRY OR BRAMBLE

*Fuchs*

**Derivation:**  Old English: *braeme, bremel*
Middle English: *brame*
**Place-names:**  Brampton, Blaeberry Gill (recorded as *Blakberygyll* in the sixteenth century), Brimbleflat (1298), Bramery (1231), Bramley, Bram Crag

The blackberry or bramble has always been a popular source of food during the few weeks in the autumn when the juicy and nutritious berries are at their best. Folklore warns that the berries should not be picked after Michaelmas Day (29 September) because it was believed that the Devil spits on them at that time and turns the fruit mouldy: the real culprit is a flesh-fly whose larvae feed on the rotting berries.

Blackberry seeds have been found in Neolithic and Viking Age graves. Medieval monks mixed blackberry juice with the juice of the mulberry as a remedy for sore throats and to treat sores and ulcers. Nicholas Culpeper notes that some condensate the juice of the leaves and the juice of the berries to keep for their use all the year. John Gerard tells us that the leaves were good for burns and swellings and that the young leaves being chewed take away inflammation of the mouth. Today the fruit of the blackberry is best known in the form of bramble jelly but John Gerard's claim that a concoction made from the leaves was used for fastening teeth back on is unknown to modern dentistry.

Bee-keepers in the Middle Ages found a practical use for the tough and pliable stems of the bramble, which were split to make the bee-skeps of those years.

The absence of any significant number of recorded place-names referring to the bramble is probably explained by the fact that it was such a common plant found everywhere that it was not considered necessary to identify particular locations.

## BOG-MYRTLE OR SWEET GALE

*Gerard*

**Derivation:**     Old English: *gagel*
                    Old Norse: *gal*
**Place-names:**   Gale, Gale Gill, Gale Bank, Galeforth, Gale
Beck, Galemire, Gale Fell, Gale Syke, Gale Garth, Gale Wood,
Gale Field, Gale Bay

The bog-myrtle is a plant of the wetlands, growing most pro-
fusely on sphagnum mosses and in damp places near to
streams and ditches and in shady woodland and lakeside
marshes. In Cumbria, as in most other northern counties, it is
commonly known as gale or sweet gale, a name derived from
Old English and Old Norse.

11

The plant is notable for its pleasant aromatic scent: Dorothy Wordsworth, writing on a warm mild morning on 16 June 1800, describes how in Little Langdale 'the valley (was) all perfumed with the Gale and wild thyme!'

The sweet fragrance of the bog-myrtle is now its chief attraction but for earlier ages it had a much more practical appeal. It was well-known as an effective insect repellent; clothing and household linen were protected from the ravages of moths if myrtle sprigs were placed among them; beds liberally endowed with the scented sprays were likely to be free from fleas, a common pest in early and medieval times.

The family broths were often given extra flavour by the addition of a few myrtle berries. The catkins or 'cones' could be boiled in water to obtain a type of wax to produce 'candleberry candles'. Equally important, and in the view of many perhaps even more important, was the unique flavour and potency imparted to ale by the addition of bog-myrtle in the brewing process. We may be sure that every housewife who was proud of her ale would know just where the best bog-myrtle was to be found.

Gale beer was once a famous northern brew, with the strength, as John Gerard noted, 'fit to make a man quickly drunke'. Indeed, bog-myrtle ale, (*porsøl* in Scandinavia), was believed to be the drink which drove the Vikings to violent and uncontrolled berserk behaviour. In contrast to this, the bog-myrtle is also traditionally associated with love, gentleness, delicacy and tenderness. In Shakespeare's *Measure for Measure* Isabella reprimands Angelo:

> *Mercifull Heaven!*
> *Thou rather with thy sharp and sulphurous bolt*
> *Spilt'st the unwedgeable and gnarled oak*
> *Than the soft myrtle.*

# BRACKEN

*Gerard*

**Derivation:**   Middle English: *braken*
Old Norse: *brakni, einstapi*
Welsh: *rhedyn*

**Place-names:** Brackenbarrow, Bracken Howe, Ainstable (1210), Brackenber (1265) – numerous examples, Bracken-lands, Glenridding (1292), Brackenburn, Brackenrigg(s) – numerous examples, Bracken Close, Brackenslack (1270), Bracken Gill, Brackenthwaite, Bracken Hill – numerous examples, Brackenwray

To many lovers of Lakeland the sweeping acres of flaming

bronze which clothe the fell-sides in autumn are one of the glories of the district. To others they represent the loss of green pastures, small farmsteads and resplendent moors of heather and bilberry. Changes in agricultural practice have undoubtedly abandoned much land to the smothering fronds and choking underground stems, but bracken is not a modern invader: most of the place-names here are first recorded in the early Middle Ages when the places where bracken grew were considered worthy of special identification. One does not have to dig very deeply into history to discover why.

Bracken was the most readily available source of winter bedding for animals and it was also harvested in some areas as a durable thatch for barns and houses. In June when bracken has a high yield of potash it was burned to produce a form of soap or lye to scour linen, then an invaluable aid in a tedious task. The Norsemen were already familiar with the special brew of ale which resulted from the addition of the uncoiled fronds of the bracken to the malt, a custom known across northern Europe from Siberia to Norway.

John Gerard relates that the root 'cast into an hogshead of wine keeps it from souring' and Nicholas Culpeper assures us that the roots boiled in mead and water produce a medicine that 'killeth both the broad and the long worms in the body', and if boiled in hog's grease the resulting ointment heals wounds and 'pricks in the body'. Both these herbalists recommend bracken smoke: Gerard encourages sciatica sufferers to 'smoke the legs thoroughly with fern bracken' and Culpeper maintains that bracken smoke drives away 'serpents, gnats and other noisome creatures which do in the night-time trouble and molest people lying in their beds'. Bracken leaves were also believed to act as a relief from nettle stings.

All these practical uses for bracken were no doubt well-known to medieval country folk. It is unlikely, however, that they were greatly impressed with the folklore tale that a concoction of bracken seed could confer invisibility. They would have agreed with the Chamberlain in Shakespeare's *Henry IV* that one was 'more beholding to the night than to fern-seed for your walking invisible'.

Bracken was one of Nature's useful and beneficial resources and it was simple common-sense to note its presence in a descriptive place-name.

# BROOM

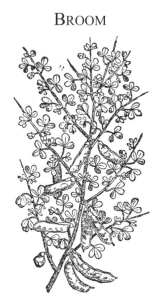

*Matthiolus*

**Derivation:**  Old English: *brom*
**Place-names:**  Bramgill, Broomfield, Bramehow (1323), Broomhill, Branthwaite, Brangull, Broomgarth, Bramley, Bromeclose, Broomlands, Bram Crag, Bromfield (1125), Broomrigg, Brampton (possible), Broom Fell, Brooms

Broom place-names are common throughout the country and for centuries it was held to be a plant of many virtues. In medieval ballads from both England and Scotland it is associated with love and magic; it also gave protection against the powers of witchcraft and the supernatural.

As the *Planta Genista* it was adopted in England and

France as the heraldic emblem of the House of Plantagenet, although this would have been of little interest to the peasant farmer of northern Britain who saw the broom rather as a source of sustenance for his winter sheep and as a protection against the dreaded 'rot'. Nor would he be likely to know that, as Gerard tells us, 'That worthie Prince of famous memorie, Henri the eight King of England, was woont to drink the distilled water of Broome flowers against surfeits and diseases therof arising', but he was probably quite familiar with Broom Tea which was – and still is – used as a mild diuretic.

We may imagine, too, that long before broom buds were eaten as an appetiser at the Coronation Feast of James II, they were well-known as a tasty morsel by many country folk who used them, as Gerard says, 'for sallads as capers be'. Broom wine featured in most recipe books until modern times but the ordinary housewife probably valued the broom more as the best of all floor sweepers. Broom is now mostly found on waste ground and by the roadside where its brilliant yellow flowers are still one of the delights of spring.

It is possible that the broom referred to in some of the place-names could be Dyers' Greenweed, a shrub related to the broom which would produce a yellow dye. The Flemish weavers who settled in Kendal in the fourteenth century soaked the cloth in the yellow dye and then dipped it in a vat of blue woad or red madder to produce the cloth worn by Shakespeare's 'three misbegotten knaves in Kendal Green.'

# THE BUTTERBUR

*Gerard*

**Derivation:**  Middle English: *burblade*
Westmorland dialect: *burblek*
**Place-names:**  Burblethwaite (1351), Burblands, Burblet
Garth

Known in various parts of the country as bog rhubarb, butter
dock, butterbur, burblek and the umbrella plant, this not particularly attractive plant of damp places by rivers, streams and
ditches was once highly valued. The huge leaves which can
grow to as much as three feet (90 cms) across are, as John Gerard described them, 'of such widenesse as that of itself it is

18

bigge and large inough to keepe a man's head from raine and from the heate of the sunne'. Indeed, the butterbur's botanical name *petasites* is derived from the Greek *petasos*, the felt hat worn by shepherds as protection from rain and heat and so often seen in statues of the god Mercury.

It was not mainly as an umbrella, however, that the large leaves of the butterbur were used in Britain. Housewives and dairymaids for more than a thousand years gathered them to wrap round their butter to keep it cool and in good condition in warm weather. The early medieval name, burblade, means literally 'butter leaf'.

The butterbur's popular name in Tudor times was the plague flower and its German name is *pestilenzwurz,* both reminders that in the days when the plague and other mysterious fevers were frequent and common afflictions, the powdered root of the butterbur taken in wine was regarded as an effective protection against this much feared pestilence. Culpeper declared that this root, 'beyond all things else, cures pestilential fevers', and Henry Lyte's *Herbal* in 1578 noted it as 'a soveraigne medicine against the plague'. It was also used as a heart tonic, driving away 'all venom and evill heate' and, in small doses, to provide relief from attacks of migraine and hay fever.

Whatever doubts we may now have about these healing powers we are bound to admire the resourcefulness of our ancestors in the versatile use they made of this strange and unprepossessing plant. The two names Burblet Garth (butterbur enclosure) and Burblethwaite (butterbur clearing) suggest that the butterbur may well have been cultivated in these places. Bee-skeps may have been placed here for the bees to gather the nectar from the early flowers which appear in March.

## CLOUDBERRY OR KNOUTBERRY AND CRANBERRY OR MARSHWORT

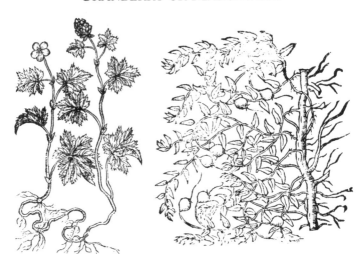

*Gerard two illustrations – Cloudberry (left) added by Johnson, 1633 and Cranberry (right).*

**Derivation:**    Old Norse: *knottr-ber* (knoutberry),
*trani-ber* (cranberry)

**Place-names:**  Knoutberry, Cranberry Hall (1770), Traneberimos (1227) now Craneberry Moss

These are plants of the peat bogs which once covered large areas of the Lake District and supported species of plant-life such as the cranberry and the cloudberry which now survive in only a few places. Drainage for agricultural expansion and periodic burning of the vegetation in the interests of better grazing have so modified the peat base and reduced the extent of the marshlands that many plants have struggled to survive these changes and the onslaught of the sheep.

So, although now rare, the cranberry and the cloudberry (or knoutberry as it is known in the northern parts of the country) are now special 'finds'. The cranberry – or craneberry – has not vanished from the Cumbrian marshlands as completely as the bird after which it was named but its pink flowers and its red autumn fruit are now found, like the orange fruits of the cloudberry, in only a few fairly remote and special localities.

The cloudberry fruits are exceptionally rich in Vitamin C and so were probably much prized by our medieval ancestors as they still are in the Scandinavian countries where they are found more abundantly and can tolerate temperatures of -30C. The ripe raspberry size berries have a taste which has been described as similar to a dry riesling wine. Cloudberries and cream may once have been a late summer luxury in Northern Britain as it still is in modern Scandinavia. It is also possible that the first Norse settlers in Cumbria may have been acquainted with the rich taste of *lakka*, the cloudberry liqueur, produced mainly in Lapland but familiar to most Scandinavians.

As a Londoner John Gerard appears to have been quite unfamiliar with the 'knotberry' and it does not figure at all in his *Great Herbal* of 1597. The omission was remedied by Thomas Johnson in his 'enlarged and amended' edition of Gerard's book in 1633.

The cranberry, too, is rich in Vitamin C and was highly valued as a nourishing addition to the autumn diet. The name 'cranberry' does not appear until the nineteenth century. Before that the wild berry was known variously as the marshwort, the mossberry, the craneberry and the fenberry. The good tart of cranberries which Queen Victoria enjoyed at Balmoral was probably of this native fruit, a version of the cranberry pie which had been relished by country folk for generations.

# DILL

*Gerard*

**Derivation:**    Old English: *dile*
                   Old Norse: *dylla*
**Place-names:**   Dillicar (1190), Dillmirebanke, Delicars

As a culinary herb and as an indispensable aid in the nursery
dill has a long and ancient history. It was as well known to
Dioscorides and the Greeks as it was to Alfric and the Anglo-
Saxons; its virtues were extolled by medieval herbalists and
by Victorian nurse-maids. As dill butter, as a pickling herb and
as a unique flavouring for soups and most meat dishes, dill ap-

23

pears alike in a thirteenth century gardening book, a Tudor *Household Book* and a twentieth century cookbook.

Dill was believed to have many medicinal uses in Anglo-Saxon times: the leechdoms of that age recommend dill as a remedy for many ailments from common stomach upsets and headaches to jaundice and loss of appetite. The discovery of dill seeds in the food remains at the Viking settlement at Jorvik and, perhaps, even the Norse origin of the word itself – Old Norse *dylla* means to lull or soothe – suggest that the faith of the Victorians in dill-water for their babies and their passion for dill-flavoured pickles were shared by their ancestors a thousand years before them.

Place-names referring to the dill from many other parts of the country appear in the Domesday Book, indicating that by 1086 the herb was well-established as a household necessity: Dillicar in Cumbria is joined by Dilworth in Lancashire, Dilham in Norfolk and Dulwich near London.

# THE MALE FERN

*Gerard*

**Derivation:** Old English: *fearn*

**Place-names:** Ferngill Crag, Ferney Green, Fernwood, Farlam (1166) = fearn + leah + ham (the homestead by the woodland clearing where ferns grow)

The medicinal powers of the male fern have been renowned for almost all recorded history. In the Classical Ancient World Pliny, Dioscorides, Galen, Paracelsus and Apuleius all wrote

25

of its virtues as accepted medical knowledge. This, together with a dash of magic, passed into the Anglo-Saxon *Lacnunga*, the eleventh century manual of contemporary medical practice, and thence into the Leech Books and Herbals of later centuries.

We may safely assume that the English settlers who came to Britain brought with them this store of folk-medicine and since, next to the female fern or bracken, the male fern is the most commonly found fern here, they had no difficulty in carrying on the tradition. They would know full well that a dose of the pulped root of the male fern was the most dramatically effective remedy for one of the scourges of their time, the stomach worm. Other milder herbs were also used for this purpose but for the dreaded tape-worm this was the only cure.

The dose was crucial to success: too large a dose could cause blindness and coma: Dioscorides advised that 'they that will use it must first eat garlicke'. John Gerard in 1597 confirmed that it was used to drive forth long flat worms from the belly and, even in modern pharmacy, until quite recently, oil capsules from the fern root were still available for the same problem.

In other respects the male fern was used in the same way as the female fern or bracken.

# FLAX

*Gerard*

**Derivation:**     Old English/Old Norse: *lín*
**Place-names:**    Limefitt, Linehams, Lind End, Linelands
(several), Linecroft, Line Riggs, Linedraw (1338), Linethwaite
(1338), Linefoot, Linewath, Linegarth, Linstock

Most of the place-names suggest that the plant referred to is
not the Purging Flax (*Linum catharticum*) or the wild Pale Flax
(*Linum bienne*) but the cultivated Flax (*Linum usitatissimum*)

much used, as its name implies, for its fibre and its seeds. Thwaite, garth, fitt, stock and croft all indicate enclosures for specific purposes, and the purpose in this case was to grow flax for food and clothing.

The oil-rich seeds, notable in the manufacture of linseed oil, were often mixed with bread flour as this was not only a palatable means of digesting them but it also prevented the bread from becoming dry. We now know that the oil is also rich in Omega 3. A mash of the seeds in boiling water makes a poultice which has a long history as an effective remedy for the relief of abscesses, sore eyes, respiratory problem, strained ligaments and sprains. The healing properties of flax were referred to in Shakespeare's *King Lear* when a servant proposes to apply it to the injury inflicted on the Duke of Cornwall:

> *I'll fetch some flax and whites of egg*
> *To apply to his bleeding face.*

The seeds were also used as an important component of cattle food, but flax achieved universal fame as the plant whose fibres could produce linen, probably the finest textile ever made. It is stronger than cotton and was in demand by the wealthy of all ages and may be seen in the exquisite weaving of much medieval work made for the elegant fashions worn at the Courts of France, England and Spain.

It was a slow and tiresome operation to separate the woody stems from the fibres which had then to be soaked and scraped and combed into the fine strands ready for the distaff and loom. For the wealthy, linen clothing was no more than they expected but for lesser mortals a new linen shirt or dress was a very special gift indeed.

Viking grave finds and medieval household accounts suggest that men of wealthy families wore linen shirts and under-breeches while women wore a linen chemise under woollen outer garments. The small yield from a crop of flax and the difficulties of transforming it into fine linen probably meant that few people possessed more than one or two garments of this precious material, and that flax may have been grown mainly for its culinary and medical uses. Even so, the many place-names, in Cumbria and elsewhere, indicate that flax was widely grown in most parts of the country.

# GRASS

*Gerard*

**Derivation:**  Old English: *gres, gaersen* = grazing land,
*beonet* (bent grass)
Old Norse: *gres*

**Place-names:**  Grasmere (1203), Bent Close, Grasmoor
Bent Holme, Grasthwaitehow (1350), Bendrigg, Grassgarth
(1305), Benty Hill, Grassgill, Bent Howe, Grassguards, Bents,
Grassenber, Bent Sike, Grassholme, Butterbent, Grassing,
Grassoms, Grassrigg

Grass was a crucially important item in the agricultural economy long before the development of the great monastic sheep runs in the early centuries of the Middle Ages. The farming patterns of the Anglo-Saxon and Norse settlers in earlier centuries were based as much on cattle-rearing as on sheep and demanded good grazing and an adequate hay harvest near to the farmsteads. The rough pastures of the uplands were more suited to their Herdwick sheep. The favoured grasses were the bents, tough, sweet upland grasses which have now been largely replaced by the sheep's fescue, while for cattle the meadow grasses were preferred and there were probably a number of species, not all of which can now be precisely identified. Gerard's illustrations of the 'small meadow grass' and the 'woolly reed grass', as he described them, may represent grasses familiar throughout medieval times.

There are few references to human use of grass but it is well-known that country maids wore bonnets made from sweet vernal grass which had the scent of new-mown hay if the wearers were caught in the rain. Of couch-grass Nicholas Culpeper comments with reference to its ancient use to treat diseases of the bladder that 'although a gardener be of another opinion, yet a physician holds half an acre of them to be worth five acres of carrots twice told over.' Whether the farmers of Grassthwaite and Grassgarth ever resorted to the grass treatment for their bladders we do not know but we may assume that the crop was primarily intended for the cows.

# HEATHER

*Fuchs*

**Derivation:**    Old English: *haeddre*
                   Old Norse: *lyng*

**Place-names:**  Hadderdale, Ling Fell, Ling Bank, Lingholm, Ling Comb, Ling How, Ling Cove (1242), Lingmell, Ling Crags, Lingmoor, Ling Croft, Lingy, Lingybank, Lings

Man has constantly changed the landscape of the Lake District but rarely more dramatically than by his intensive exploitation of the uplands in the interests of sheep farming. In Cumbria, as in Sir Thomas More's *Utopia*, sheep were once only 'small eaters' but they have increasingly become 'great devourers', so much so that their voracious grazing has transformed the landscape of the fells. The consequence for much of the fell-flora was devastating but modern conservation methods are now having an effect.

With the clearance of the great medieval forests, the lower fell-sides, left to themselves, would have been (and in many places were) covered with a wealth of heather and other mountain flora such as are now found in Scotland and in the alpine meadows of Switzerland. Forest clearance and overgrazing, followed by erosion, changed all this, and where there were once vistas of heather, bilberry and a host of flowers, we now see fell-sides clothed in bracken and nardus grass. It is still possible to find, as on Black Combe and on the great sweep of the moors near Alston, an impressive display of autumn purples but within the National Park in Lakeland most of the great heather moors were lost as the grouse shoots were abandoned and the sheep moved in.

The disappearance of the heather is, for us, mainly a scenic loss but to earlier generations it would have meant a great deal more. For, to them, heather was almost a necessity of life. The beds on which they slept were often made of it as was the roofing thatch which kept them dry; the besoms which swept house and byre were fashioned from it; the fires which burned on their hearths were kept alive and bright with it.

Heather, too, was a valuable source of nectar for bees. Honey which is now known to have antibiotic, antibacterial and anti-inflammatory properties, and such was their simple but comprehensive knowledge of herbal medicine that our medieval ancestors may well have been thoroughly familiar with these virtues. Heather was the chief source of the honey they used for sweetening and, especially, for the brewing of mead, the legendary drink of Picts, Anglo-Saxons and Vikings.

Heather also added that special flavour to the family ale and its tough wiry stems made excellent baskets, ropes, hooks and other useful gadgets for farm and house. The young shoots provided late winter feed for lambing ewes while the abundant grouse which made the heather moors their home were always good for the tables of those who were astute enough to take them. The folk who gave names to Lingmoor and Lingmell may have been just as appreciative as we are today of the purple vistas and heady scents of the heather but they also viewed it with an eye to its practical uses.

# HOUND'S TONGUE

*Gerard*

**Derivation:**  Old English: *ribbe*
**Place-names:**  Ribton (twelfth century), Rib End

A herb called *ribbe* appears as one of the 'magic medicines' in the Anglo-Saxon Herbal and this is now generally accepted as a reference to the plant we know as the Hound's Tongue and not to the Ribwort which is a plantain and quite a different plant.

The Hound's Tongue has one peculiar characteristic which gives it a certain distinction. This is its rather strong and unpleasant mousey smell, designed it was once believed to repel dogs, a belief also associated with the shape and soft texture of the leaves. John Gerard confidently asserted that 'It will tye the tongues of houndes so that they shall not bark at you if it be laid under the bottom of your feet,' a belief affirmed by Nicholas Culpeper who also claims that he cured the biting of a mad dog using only this medicine. More practically, perhaps, we are told that the leaves boiled in hog's grease produced an ointment to treat scalds, burns and skin diseases; and Nicholas Culpeper adds, for good measure, that 'the baked roots are good for piles.'

Against all this, the Hound's Tongue has another more pleasant feature, well-known to generations of children in former days. At the base of the purple funnel-shaped flowers is a supply of sweet nectar, a delight to be sucked out and enjoyed. It is not at all a common plant and we are fortunate that a Cumbrian place-name has survived since the twelfth century to remind us of the mixture of medicine and magic which surrounded plants and herbs for so many generations.

The virtues of the hound's tongue were probably widely known in medieval England as the name also survives in places as far apart as Ribbesford in Hereford and Worcester, and Ribston in North Yorkshire.

# MADDER

*Fuchs*

**Derivation:**   Old English: *mæddre*
                  Old Norse: *mathra*
**Place-names:**  Mattergill, Matterdale (1250)

This is the true madder, *Rubia tinctorum*, which, like woad, was cultivated in Britain possibly from Celtic times, chiefly

37

for the red dye obtained from its root. The cultivated plant produced a much better dye than the wild madder which was of interest more for its tangle of prickly stems which were used to polish metal, a kind of natural 'steel-wool'. The collection of textile pieces recovered from the excavations at Jorvik have enabled archaeologists to confirm for the first time what Chaucer told us six hundred years ago, that madder, weld and woad were the three principal plants used by medieval dyers to obtain the reds, yellows and blues which were the basic colours of their craft.

Place-names in Cumbria do not appear to record the cultivation of either weld or woad but Madersdale (or Mathersdal) would seem to establish that madder had been grown there for some time before the thirteenth century. The famous Redcoats of the British army were dyed with the red of the madder. Cultivation of all these plants for their natural dyes virtually ceased in Europe after the discovery in the nineteenth century of methods of producing chemical dyes. Madder was also believed to be helpful in removing blemishes from the skin such as freckles. Culpeper wrote, rather obscurely, 'that a herbal concoction of madder hath an opening quality and afterwards to bind and strengthen.'

# MARSH MARIGOLD

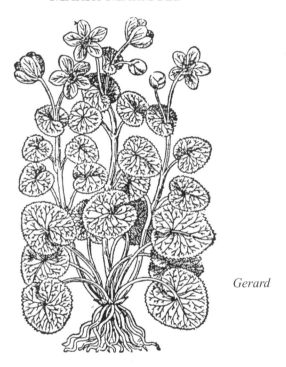

*Gerard*

**Derivation:**   Old English: *golde*
**Place-names:**   Goldrill (1573), Gouldmer, Gouldmire, Gowan (River)

The shining golden flowers and brilliant green leaves of the Marsh Marigold have brightened the days of spring and early summer in Northern Europe since the last Ice Age. It is believed to be one of the few native plants to have survived the Ice Age and they are now so universal and abundant that the flowers have acquired a different popular name in every European country and, in Britain, in almost every county.

Its northern names include Kingcup, Mayflower, Mayblob and Gowan. Dorothy Wordsworth, walking in Easedale, described how 'the Gowans were flourishing along the Banks of the stream.' Most of the place-names are derived from the Old English *golde*, to which the medieval church added an association with the Virgin Mary (Marigold) to adapt the persistent pagan custom of strewing the flowers in and near cottages on May Day to ward off the powers of evil and witchcraft.

The plant seems to have had no great medicinal reputation. The Anglo-Saxon Herbal gives it a passing reference as a remedy for minor skin rashes and it was believed that the exhalations from the flowers could help in the treatment of fits, but since many parts of the plant are an irritant or even poisonous it was not held in any esteem as a medicine. John Gerard dismisses its virtues brusquely: 'touching the faculties of these plants we have nothing to saie.'

He does go on, however, to give a vivid description of the plant: 'Marsh Marigold hath great broad leaves, somewhat round, smooth, of a gallant green colour, slightly indented or purld about the edges, among which rise up thicke fat stalkes, likewise green; whereupon do grow goodly yellow flowers, glittering like gold.' Shakespeare put it more poetically when he wrote of the joys of a spring dawn: the lark singing at heaven's gate, the dew lying in the 'chaliced flowers' and the 'winking mary-buds' opening 'their golden eyes'.

These are sentiments more easily expressed in the more comfortable, prosperous world of Elizabethan England than in the hard and unromantic years of the tenth century, years of breaking the soil and stubbing the waste. The marsh marigold does not appear in Cumbrian place-names until the sixteenth century.

# NETTLE

*Fuchs*

**Derivation:**     Old English: *netele*
Old Norse: *nata*

**Place-names:**   Nateby (1242) = the farmstead where nettles grow or Nati's farmstead, Nettle Gill, Nettle Pot; Nategill, Nettle Slack, Natland = the grove where nettles grow or the grove belonging to Nati

Now mainly a plant of the woods, hedgerows and abandoned ruins, and held in little regard, the nettle was once highly esteemed and even extensively cultivated. No other green plant is so rich in its mineral and vitamin content. 'It hath a kind of cleansing qualitie,' declared John Gerard, and modern herbalists would entirely agree with him. Until the eighteenth century when other vegetables became more commonly available, young nettle tops, boiled lightly to remove the sting and mixed with oats and butter, were well-known as an excellent vegetable.

In Walter Scott's *Rob Roy* the gardener at Lochleven raised early nettles under glass for this purpose. Samuel Pepys records in his Diary for February 1661: 'We did eat some nettle porridge, which was very good,' a recipe of nettles, leeks, broccoli, rice, salt, pepper and melted butter. Nettle soup, nettle tea and nettle beer were also once well-known items in the household diet. The juice of the nettle was used in the kitchen as a substitute for rennet in cheese making and a bunch of nettles hanging in the larder helped to keep flies away.

Among the many bizarre methods people have thought of to mortify the flesh must surely be numbered that known as 'urtication', beating the body with the intensely stinging stems of *urtica dioica* or the stinging nettle. As a means to spiritual enhancement this was said to be the instrument of choice among those who had taken monastic vows, but flagellation in this way served a double purpose, for by that time in human history it was common knowledge that this was believed to be an effective treatment for rheumatism and arthritis as it created body heat and stimulated the circulation of the blood. The Romans believed firmly in this and cultivated the nettle for this purpose.

Probably older even than the nettle prescription for rheumatic pains is its use in the manufacture of nettle cloth. Nettle fabrics have been found in Bronze Age and Viking graves and there is evidence that it was cultivated for this purpose in medieval Norway; nettle table-cloths and bed-sheets were made in eighteenth century Scotland while nettle weaving continued in Silesia and the Tyrol until the twentieth century and continues to this day in some Himalayan countries.

During the First World War nettles were widely grown in Germany to make army uniforms. Nettle fabric was even considered fit for a princess. In Hans Andersen's tale of *The Princess and the Swans* we are told that the coat made by the princess was woven from the nettle. We may imagine, too, that she may have washed her hair in nettle shampoo, a concoction made from the leaves, flowers and seeds, good for the scalp and enhancing greatly the colour and texture of the hair. The peasant farmer saw the nettle in a less romantic light: for him it was a nourishing food for his cattle and pigs, and fed chopped and dried to his poultry it increased egg production.

In view of the nettle's many virtues it is not at all surprising that in an earlier age it should have been so highly regarded and even so carefully cultivated.

# RAMSONS OR WILD GARLIC

*Gerard (illustration
by Johnson)*

**Derivation:** Old English: *hramsa*
**Place-names:** Rampsbeck, Rampson, Rampsgill, Rampholme, Rampshaw

Ramsons flourish in moist soils and most commonly in ancient woodlands where, in the early spring, before the leaves on the trees shut out the sunlight, carpets of its star-like white flowers and elegant green leaves cover the ground and fill the air with the scent of garlic. This is often accompanied by a display of bluebells to make a fine spring panorama of natural beauty. Apart from this brief attraction ramsons are now largely ignored – or even shunned because of their strong garlic aroma and their close resemblance to the poisonous lily of the valley – but to earlier generations they were held in high regard.

All parts of the plant – leaves, flowers and root bulbs – are edible and were used in salads and soups, boiled as a vegetable, made into garlic butter, cheese and bread, and used in the way that we now use chives.

As a healthy addition to the family diet after the restrictions of the long winter months this restored many ailing folk to well-being, for wild garlic was a most effective agent in cleansing and purifying the blood. It was also found to have benefits in the treatment of ailments such as asthma and other respiratory problems and was widely used to treat whooping cough and digestive upsets. A herbal concoction from the boiled leaves could be used as an antiseptic to treat cuts and wounds and as a disinfectant.

Modern medical research is largely inconclusive in its verdict on many of the virtues once attributed to ramsons or even to the stronger bulbs of modern garlic, but we should not lightly dismiss the faith in the efficacy of the wild garlic in which our ancestors clearly believed.

The norsemen of the Viking age were especially fond of garlic in their diet and one must assume that they were socially less sensitive than the theatre-goers of Shakespeare's day, for in *A Midsummer Night's Dream* Bottom appeals to his 'actors' to

'eat no onions nor garlic, for we are to utter sweet breath, and I do not doubt but to hear them say, it is a sweet company.'

## REEDS, RUSHES AND SEDGES

*Gerard - left, common rush and right common reed.*

**Derivation:**  Old English: *hreod* = reed, *rise* = rush,
*secg* = sedge.
Old Norse: *sef* = sedge, *storr* = star sedge

**Place-names:**

REED – Readmire, Reathwaite (1265), Redekar (1241), Red-
mire, Reed Moss (1241), Redhow, Redness, Redsyke

RUSH – Rash, Rish, Wellrash

SEDGE – Seathwaite (1292), Seavy Holm, Seavy Side,
Seavysyke (1303), Segdale, Seavyrigg, Seggs

STAR SEDGE – Star Crag, Stare Beck, Starfitts, Stargill (1230),
Starnmire (1285)

These well-known plants of the wetlands were once much more common than they are today. Drainage of the valleys and lower fell-sides to create the green and tidy pastures of modern times severely restricted the areas where, in earlier days, reed, rush and sedge had found a congenial habitat. These marshy places were not then looked upon as 'wasted' farmland but as sources of plentiful game – or even fish – and of abundant materials for baskets, mats and frails, even wall-hangings.

These tough plants were also a valuable thatch for cottages and barns. The strong leaves of the sedge were harvested for the special purpose of forming the ridge along the top of the reed-thatched roofs and in northern counties this was known as thack; hence we have place-names such as Thackmire, Thackthwaite, Thack Dales and Thack Bolton, and names such as Thackeray. Shakespeare refers to reed-thatch in *The Tempest*:

> 'His tears ran down his beard, like winter's drops
> from eaves of reeds'

The pith of the rush, dipped in tallow, was essential in making rush-light candles when the rush-candle was used in every country cottage. In Cumbria the rushes preferred for this purpose were known as sieves (from the Norse word *sef*), the word used in William Dickinson's dialect account of the autumn harvesting for rush-lights:

> T'young fwoks'll gang
> Till a cannel-seave syke,
> And pick a shaff shangans for leets;
> Than hotter to heamm through bog and wet dyke,
> To peel them and dip them at neets.

Each rush-light burned for about 30 minutes and so every

household might need as many as 7-8,000 rush-lights to last the winter months.

Every church and dwelling house strewed its floors with fresh rushes from time to time to keep them clean and sweet A guest whose room was not freshly strewn knew that he was 'not worth a rush'!

The many practical uses of reeds, rushes and sedges would make the places where they grew noteworthy but the rush, especially, was especially esteemed as the principal feature of the ancient rush-bearing ceremony, a semi-religious ritual, probably dating from pagan times, which was accompanied by a good deal of merry-making. This renewal of the rushes on the church floor, once an important necessity, was no longer needed when the floors were paved but a form of the annual ritual is still held in a number of Cumbrian villages.

# SORREL

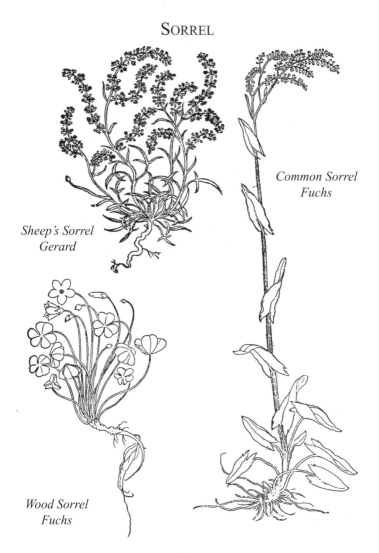

*Common Sorrel*
*Fuchs*

*Sheep's Sorrel*
*Gerard*

*Wood Sorrel*
*Fuchs*

51

**Derivation:**    Old English: *docce*
                      Middle English: *sorrell*
**Place-names:**  Sorrellsyke, Docker, Dockpull (1352), Dockray

The precise meaning of some of these place-names is unclear. The derivation may be from the Old Norse *døkk*, a hollow, and if the correct origin is Old English *docce* it is not possible to say which of the three sorrells is referred to beyond the fact that sheep's sorrell prefers a dry situation. The 'dock' of Dock Tarn is almost certainly the water lily.

The common sorrel, the wood sorrel and the sheep's sorrel all share several popular names – bread and cheese, cuckoo bread, green sauce and sour dock. This reflects the fact that they all have similar medicinal properties and culinary usefulness, and to the housewife gathering the leaves for a spring salad the distinction between them would have been of little interest.

The botanical names for these three types of sorrel provide the key to the great enthusiasm shown in earlier days for sorrel leaves. *Rumex acetosa* (common sorrell), *oxalis acetosella* (wood sorrell), and *rumex acetosella* (sheep's sorrell), all contain calcium oxalate, a salt of oxalic acid, which gives the leaves a sharp taste admirable for flavouring in much the same way as we use lemon juice. *Rumex* is derived from the Latin *rumo*, to chew or suck, a name acquired from the ancient country custom of chewing sorrel leaves to assuage thirst and the tubers of wood sorrel as a rather desperate source of food or medicine.

A much later age discovered that this same oxalic acid could be used to remove iron stains or ink from clothing or linen, a household hint familiar to every laundry maid in Victorian times. The early Herbals are eloquent in their praises of sorrel roots and leaves in the treatment of scurvy and afflictions of the kidneys or liver, as an eye lotion and as a cooling agent in cases of high fever. An Elizabethan cookbook even had a sorrel recipe, which, it was claimed, would assist fertility by strengthening the seed of both man and woman.

However it was as a salad served with butter or as a green sauce to garnish pork or goose that sorrel was held in high esteem: 'Of all the sauces,' wrote John Gerard, 'sorrel is the best, not only in virtue but also in pleasantness of taste.' A fifteenth century poem on gardens reveals that sorrel was commonly cultivated for these purposes and, two hundred years later, John Evelyn included it among his list of *Plants for the Kitchen-Garden*. All the place-names from Cumbria date from the thirteenth or fourteenth century in their earliest recorded form, an indication that, long before it graced the royal table at the Court of Henry VIII, the sorrel and its virtues were known to ordinary cottage folk everywhere.

# STRAWBERRY

*Fuchs*

**Derivation:**  Old English: *streaw-berige*
**Place-names:**  Strawberry Bank (several), Strawberry How, Strawberry Wood

None of the place-names is recorded before the reign of Elizabeth I and the strawberry they refer to is *Fragaria Vesca*, the small, wild strawberry. The large modern cultivated fruit was first raised in France in the eighteenth century but there is abundant evidence that strawberries were very well-known long before this. The name itself is Old English and it appears in a tenth century Anglo-Saxon plant list.

By 1265 the household accounts of the Countess of Leicester indicate that the 'straberie' was a familiar dessert and less than two hundred years later 'Strabery ripe' had a secure place among the famous street-cries of London. Cardinal Wolsey in the reign of Henry VIII was famously partial to a dish of strawberries and cream, a dish which was to become the festive hallmark of the English summer. Even in the midst of political plots and conspiracies Shakespeare's Duke of Gloucester found time to beg some strawberries from the Bishop of Ely: 'My Lord of Ely, when I was last in Holborn I saw good strawberries in your garden there.'

The good bishop seems to have known a great deal about the strawberry for he uses it to illustrate his philosophy of human nature in a comparison which would probably not appeal to the modern gardener:

> *The strawberry grows underneath the nettle,*
> *And wholesome berries thrive and ripen best*
> *Neighbour'd by fruit of baser quality.*

Everyone in Tudor England would have appreciated Bishop Latimer's description of that blot on the Tudor Church, the non-resident parish priest, as 'strawberry preachers' who 'come but once a year and tarry not long'. Everyone shopping in the local market would have come across Queen Elizabeth's 'strawberry wives that laid two or three good strawberries at the mouth of their pot and all the rest were little ones.'

The fresh wild strawberry with its 'pleasant, tart, grateful taste, and an agreeable smell,' as Culpeper put it, had been gathered and enjoyed by rich and poor alike for at least a thousand years before the horticulturalists introduced the new varieties. And most of them would probably have agreed with Izaak Walton's Dr Boteler that: 'Doubtless God could have made a better berry, but doubtless God never did.'

The juice of the strawberry was used to bathe inflamed eyes and as a gargle for a sore throat. Culpeper also claims that 'the water of the berries, carefully distilled, is a sovereign remedy and cordial in the panting and beating of the heart, and is good for the yellow jaundice.' Today it is presented as a fruit rich in Vitamin C and as an aid to reducing one's level of cholesterol.

# TANSY

*Gerard*

**Derivation:**    Middle English: *tanesie*
**Place-names:**  Tansy Garth, Tansy Gill

This now sadly neglected herb with its sharp, spicy aroma and its peppery taste was once one of the most highly valued of our native plants. Its dark, fern-like leaves and its clusters of

57

yellow flower discs still provide a distinctive splash of colour to the late summer hedgerows but only rarely now is it especially grown for use in cooking and perhaps not at all for its use in medicine.

The tansy was once held in such esteem for both these purposes that the Norsemen took it with them to the Faeroe Islands and the Pilgrim Fathers carried it over to New England. Tansy had so many medicinal virtues that it was looked upon as a panacea for almost all ills but it seems to have been particularly effective in the treatment of worms and stomach disorders, complaints of the bladder and kidneys, blood disorders and problems of circulation, fevers, and external swellings and bruises. A fourteenth century Herbal refers to Tansy Tea as a treatment for feverish illnesses and to the leaves as a healing agent for wounds. In the sixteenth century it was considered to be a necessity for every garden and John Gerard maintained that it was especially useful against the gout.

The young spring foliage was eaten as a salad and restorative after the restricted diet of the late winter months – as Culpeper put it, 'to consume the phlegmatic humours which the cold and moist constitution of winter most usually affects the body of man with.' Tansy puddings and tansy Easter cake and tansy omelettes became traditional fare at this time of the year and the dried and powdered flowers were kept to add to flour to give colour and flavour to tansy buns. Another attribute of this versatile and strongly aromatic herb was its efficacy in repelling insects and mice from food stores and bed linen.

Most of these household, medicinal and culinary uses of the tansy were passed down the generations and were still part of the lore of the Victorian age but it is doubtful if the ancient belief that it could be an aid to achieve conception survived quite so close to modern times. However it was less than a few generations earlier that Nicholas Culpeper had exhorted 'women that desire children' to 'love this herb, it is their best companion, their husband excepted.'

We have no means of establishing the extent to which the tansy reinforced their husbands' efforts in ensuring conception but it was altogether a risky enterprise since too much of the oil of tansy could also result in miscarriage and abortion. It was actually used to induce abortion when this was considered necessary or desired. An overdose could also end in death. Tansy was associated with death and was widely used at funerals and also enclosed in coffins to preserve bodies from corruption. This may have had something to do with the Greek origin of the name – *Athanasia* or immortality.

# THISTLE

*Cotton Thistle
Fuchs*

**Derivation:**    Old English: *thistel*
Old Norse: *thistill*
**Place-names:**    Thistlereddings, Thistlerigg, Thistillehalfaker
(1280), Thistlebar, Thistleton, Thistlebarrow, Thistlethwaite,
Thistlefield, Thistlewood, Beckthistle

More than a dozen varieties of thistle may be found in Britain and it is a matter of some speculation as to which is referred to in the many Cumbrian place-names where this plant appears. The most common thistles found in Cumbria today are the spear thistle, the marsh thistle and the creeping thistle. But this does not necessarily tell us which of the many species of thistle attracted the attention, favourable or otherwise, of the medieval farmer struggling to grow his crops, or of his wife seeking to provide the necessities of the household.

To the farmer all thistles were a menace. After all, the Book of Genesis relates that after the Fall in the Garden of Eden, God cursed the ground and threatened Adam that only 'thorns and thistle shall it bring forth to thee'. The thistle had always been the enemy of cultivation.

Shakespeare's Duke of Burgundy knew that in neglected fields nothing teems:

> *But hateful docks, rough thistles, kecksies, burs,*
> *Losing both beauty and utility.*

The common field or creeping thistle has always been a major problem and it is not at all improbable that some, at least, of our place-names refer to these. They are of no use to the housewife, however. For her it was the cotton thistle and the spear thistle which had certain attractions. For stuffing pillows she could not gather too much of the soft, fluffy down while the seeds produced an oil which was invaluable in cooking, altogether fresher and tastier than animal fat.

The young stems could be boiled, 'to disarm them of their prickles', as John Evelyn informs us – and served with butter as a salad or vegetable. Nicholas Culpeper, writing in the mid-seventeenth century, noted that Galen, the second century Greek physician who throughout the Middle Ages was still regarded as the supreme authority on medicine, stated that the roots and leaves of the cotton thistle were valuable in the treatment of rickets in children and also good for such persons who suffer from spasms or convulsions or a crick in the neck.

Whether the names Thistlethwaite (thistle clearing) or the thirteenth century Thistillehalfaker (a cultivated area of thistles) suggest that these useful thistles were actually grown as a crop, is open to speculation, but the idea compels one to admire the resourcefulness of a simple folk always seeking to improve their hard life with the materials nature had to offer them.

# WATERCRESS

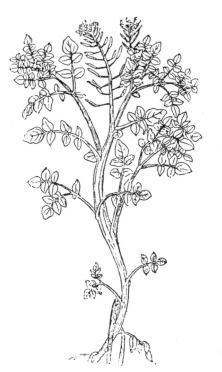

*Fuchs*

**Derivation:**  Old English: *cærse, wielle-cærse*
**Place-names:**  Caskew, Carswelhowe (1523 – now How Hall), Creskeld (1200), Cresskeld – several examples, Kershope (1200), Kerseyflatts

The green, tangy leaves of watercress, growing by fresh flowing springs were a source of valuable, vitamin-rich salad food in the difficult days of the late winter when the abundance of late summer and autumn was only a distant memory. The dull and restricted diet of the long months after the Christmas festivities began to take its toll as the days lengthened. Blood, skin and appetite yearned for fresh greens to remedy what today we would call a deficiency in Vitamin C in which watercress is particularly rich.

The tonic qualities of watercress, celebrated in medieval poetry, were effective against scurvy, anaemia, blood disorders and even that scourge of medieval families the stomach-worm, all strong incentives to make a special note of the springs where it grew. But for some, no doubt, there could be no stronger incentive to seek out and eat watercress than that mentioned by John Gerard who assures us that 'being chopped and boiled in wine or milke... it doth cure maidens of the green sicknesse... and sendeth into their faces their accustomed lively colour,' a prescription confirmed rather less elegantly by Nicholas Culpeper who brusquely comments that 'the leaves bruised or the juice will free the face from blotches, spots and blemishes.'

The watercress plants grown commercially today are almost identical to those found in the wild and eaten by our ancestors who did not have to concern themselves with the effects of agricultural chemicals and probably knew nothing of the liver fluke and its infestation of the waters and pastures where untreated sheep are allowed to graze.

The place-names 'Cressing' in Essex and 'Cressingham' in Norfolk and the numerous names 'Cresswell' and 'Creswell' found in Northumberland, Derbyshire and Staffordshire indicate that the value of this plant was widely known in medieval England.

## WILD ROSE (Briar, Dog Rose or Choop)

*Zwingler*

**Derivation:** Old English: *brer*, *heope* (Heope is the origin of the Cumbrian dialect name choop)

**Place-names:** Choup Gill, Brearflate, Chowpow, Breary Slack, Shooptree Nook, Breerhow, Briers, Briery (1283), Briery Close, Bryers, Bruthwaite

Place-names referring to the wild rose are widely distributed in Cumbria, occurring most frequently, as one would expect, in field-names. The briar rose is most commonly used in these names but the northern dialect name choop is also occasionally found, derived from the Old English *heope*.

In Ancient Greece the wild rose appears in Homer's *Iliad* as the flower of love, and with its contrasting features of delicate perfume, sharp thorns and blushing pink, it has long been the poetic symbol of the pleasures and pains of youthful rapture. In more recent times it was extolled by the poet Rupert Brooke as the epitome of the treasured eccentricity of the Englishman's freedom:

> *Unkempt about those hedges blows*
> *An English unofficial rose*

The inhabitants of medieval Britain cast a more practical eye on the virtues of the wild rose than either Rupert Brooke or Homer but just as Aphrodite knew that oil of roses would cure Hector's wounds so they, too, were aware of the valuable tonic properties hidden in the bright, shining berries. A mead fermented with rose hips was an exhilarating concoction of great nutritious value while rose-hip tarts and pies made a tasty dish brimming with 'nutritious' vitamins.

John Gerard's *Grete Herball* at the end of Queen Elizabeth's reign, tells us that 'tartes' of the ripe fruit were used for 'banketting dishes', suggesting that it was not only in the humble cottage that these were appreciated. Even today we are familiar with these 'tartes' and also with rose-hip tea, a beverage made from the dried hips infused in boiling water.

The hips have also for many generations been used in rose-hip soup and rose-hip preserves and various concoctions for herbal medicines. The berries are exceptionally rich in Vitamin C and so have numerous health benefits while their high Vitamin A content has long been known to be beneficial in the treatment of skin troubles – a common health problem in earlier times.

Less authenticated was the long-held belief that the roots of the dog-rose could cure infected wounds caused by the bites of a wolf or mad dog, an ancient superstition perpetuated by Pliny who told a story of a soldier of the Praetorian Guard who, it was said, was cured of rabies in this way. Medieval apothecaries added their own dubious piece of folklore by peddling a 'medicine' made from the galls which grow on the wild rose and are often known as bedeguars or Robin's pincushions. Culpeper faithfully records the prescription: 'In the middle of the Balls are often found certain white worms, which being dried and made into powder, and some of it drunk, is found by Experience of many, to kill and drive forth the Worms of the Belly.'

More cheerfully, John Gerard relates that 'Children with great delight eat the berries, make chains and other pretty gewgaws of the fruite.' Children through all the centuries have had good reason to ask:

> *Do you know where the wild roses grow*
> *So sweet and scarlet and free?*

# 2
## CEREALS AND PULSES, ETC

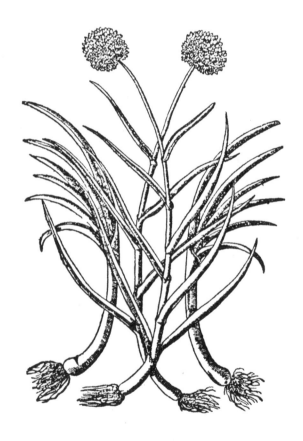

The Norsemen, and before them the Angles and Saxons, who came to settle in Cumbria were mostly farmers rather than marauding Vikings, and it should, therefore, come as no great surprise that archaeological and place-name evidence from both Britain and Scandinavia points clearly to the cultivation of significant quantities of cereals and pulses. Yet the evidence of the excavations at the Viking urban settlement at Jorvik in York seems to indicate that the diet of the townsfolk apparently consisted of a high proportion of protein with the meat of cattle, sheep, deer, goats and pigs forming a large part of their daily fare.

Barley was the main cereal crop not only because it is particularly suited to a northern climate and poor soils but also because it was of major importance as a source of food and

*Wheat*        *Matthiolus*        *Barley*

drink – flour for bread, malt for ale and as the chief ingredient of the porridge or gruel which was a regular part of the daily diet. Rye and oats were also grown for flour in Cumbria, as they were throughout northern Europe, but wheat was at first less widely known. The Norse Eddic poem, *Rigstula*, refers to 'the coarse loaf, heavy and thick, stuffed with bran' and to 'the thin loaves, white, made of wheat,' the latter clearly of better quality flour and more finely ground, a luxury.

The usual method of baking bread, known as 'haver bread' – from the Old Norse word *hafri* meaning oats – was to place the dough on a flat circular plate with a long handle which was rested on a flat stone covering the hot embers of the fire. This was the 'bakestone' which appears in many Cumbrian place-names as in Bakestones, Backstones, Baxton Holme and Baxton Gill, places where these stones were found.

Peas were extensively cultivated, mainly for the nutritious pease pudding which featured so prominently in the countryman's diet throughout the Middle Ages. Beans, too, were widely grown and, like peas, were often eaten raw, but they were also one of the important ingredients in the ever-simmering cauldron of *grautr* or thick soup which stood, with its great wooden ladle alongside, on every hearth stone.

An angry comment by St. Olaf, provoked by the political ambitions of King Knut, 'Does he mean to eat up all the cabbage in England?' suggests that a form of this vegetable was commonly grown at that time, but there are very few place-names in Cumbria, or elsewhere in England, which may refer to the cabbage.

Leeks and garlic, on the other hand, were very well-known and formed a substantial part of the regular diet, a custom which must have created its own special ambience in family and social intercourse. Gerard tells us that the leaves were eaten with butter by such as are of stern constitution but he was writing in the less austere days of the late sixteenth century whereas in Viking times both were eaten raw. The medicinal virtues of leeks and garlic were universally recognised. As an anti-septic and anti-toxin they had no rival in herbal medicine and they were believed to be especially effective as a remedy for cleansing the blood, curing skin troubles and expelling worms.

One of the earliest references to the leek or garlic occurs in Snorre Sturlason's account of the death of Tormod Kolbrunarskald after the battle of Stiklestad in 1030. Severely

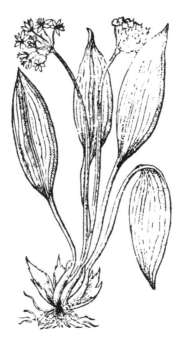

*Gerard*

wounded in the battle, Tormod went to a nearby barn where a woman 'doctor' attended to the wounded. She devised a crude but effective diagnosis to ascertain the extent of their injuries. A porridge of boiled leeks or garlic and other herbs was administered and if a pungent smell later came from the wound this was a sure indication that the intestines had been pierced and the man would surely die of peritonitis, so no further action was taken.

It is not always possible to make a distinction between leek and garlic in interpreting early place-names. Just as in our modern botanical classification all belong to the *allium* family so in the Norse language the word *laukr* appears to refer to both leek and garlic.

Ramsons, which is also an *allium* and has the distinct smell and flavour of garlic, was known to the Norse as *hramsa* and its leaves were used to add a strong flavour to the *grautr* or porridge. It still grows commonly in any moist place and its unmistakeable smell and not unhandsome white flowers may be found by almost any shady beck or in any damp woodland.

**Barley**    Old English: *bere*
              Old Norse: *bygg*
**Place-names:**  Barton, Bigland, Barwise, Biggersbank, Barwick, Bigrigg, Berslack, Bigg Close, Barrackbank, Biggards

**Rye**    Old English: *rygg*
           Old Norse: *rugr*
**Place-names:**  Rydal, Ruckcroft, Rycroft, Ryegarth, Ryelands

**Oats**    Old English: *ate*
            Old Norse: *hafri*
**Place-names:**  Blachateridding (1200) = the clearing for black oats, Haverthwaite, Haverigg, Haverlands, Haverbrack, Haverflatts, Haveridding

**Wheat**    Old English: *whæte*
             Old Norse: *hveit*
**Place-names:**  Waytslack, Wheatriddin, Waytcroft, Waytlyth

**Peas**    Old English: *pise*
**Place-names:**  Peaseland, Peasehow, Peaseber, Peasemire, Peasedales

**Beans**    Old English: *bean*
             Old Norse: *baun*
**Place-names:**  Beanlands, Bounecroft, Beanthwaite, Beanwray

**Cabbage**       Old English: *cal*
                  Old Norse: *kál*
**Place-names:**  Calecroft, Cailebanks, Calebreck

**Leeks**         Old English: *leac*
                  Old Norse: *laukr*
**Place-names:**  Larkrigg, Lockholme, Lacra, Loughrigg (St Bees)

**Ramsons**       Old Norse: *hramsa*
**Place-names:**  Rampsgill, Rampsbeck, Rampsholme, Rampshaw, Rampson

# 3
## TREES

# ALDER TREE

*Matthiolus*

**Derivation:**  Old Norse: *elri*
Middle English: *eller*

**Place-names:** Aldersceugh, Ellerside, Ellers (several), Ellerholme, Eller Beck (several), Elliker, Elleray, Eller How, Ellercar, Ellershaw, Eller Dubs, Ellerthwaite, Eller Gill (1310), Seatoller (?)

The alder is a familiar tree in almost every part of Cumbria. Its little cone-like fruits, dark and woody, remain on the tree throughout the winter, and its ripening catkins, male and female together, delight the days of early spring. Now found mainly along the banks of rivers and woodland streams, on the shores of lakes, and in damp, low-lying places, the alder once covered the undrained valley floors and the swampy slopes as high as about 1,000 feet. One of the first tasks of the early settlers would have been to drain off the surface water and clear the ground of alder and scrub: only then could they begin to create fields for arable and pasture.

The smoke rising from the fast-burning bonfires of alder must have drifted over the newly-built homesteads during many years of unremitting toil. Even when the work of a lifetime had produced the first lush meadows and acres of barley and rye, there were still vast 'carrs' of boggy woodland and moss where alder grew so thickly that such places could be identified by name – Ellers, the place overgrown with alder; Ellercar, the fenny land full of alders; Aldersceugh, the alder wood, Ellerside, the hill-slope where alders grow.

Later ages added more names when it was discovered that alder wood made good charcoal for the manufacture of gunpowder and was the best material for clogs and shoe-soles, for water-troughs and spinning wheels. The earliest-known causeways or footways over wetlands were constructed from water-resistant alder wood. Salmon smoked in slow-burning alder wood was found to acquire an especially good flavour.

The alder was well-known as the source of both a black dye and a green dye. The dyers found the bark of alder useful, as Gerard informs us in the sixteenth century, for a black dye 'much used of poor countrie diers for the dying of course cloth, caps, hose and such like.' The flowers produced a green dye said to have been used for the famous green dress worn by the fairies and by forest outlaws such as Robin Hood who wished to fade into the background of the greenwood.

There is no evidence whatever to suggest that the English and Norse folk who came to Britain were deterred from felling, burning and making full use of the alder by the old Teutonic legend of the evil spirit, the Erlkönig or Alder King,* whose wrath fell upon those who attacked him, and they were probably unaware that it was a sacred tree to the Druids.

The dyers, charcoal burners and gunpowder makers wrought havoc in the alder woods, but they have now departed and the alder, unloved by deer or sheep, is once again becoming one of the most common of northern trees.

---

* This name and legend are derived from a German mistranslation of the Danish *elverkonge* meaning King of the Elves.

# APPLE TREE

*Matthiolus*

**Derivation:** Old English: *æppel*
Old Norse: *epli*
Middle English: *crabbe*

**Place-names:**   Appleby, Appletreeholm, Applegarth, Apple-treeworth, Appletreekelde (1250), Applethwaite, Appleton, Crabtree (1597), Appletree, Crabbank (1485), Crabtree Side (1655), Appletreebank, Crabstack

The cultivated apple as we know it today in all its many varieties dates from the experiments of sixteenth century horticulturalists with the native crab apple and the red-skinned pippins imported from France. Before that the crab apple had no rivals other than the strangely shaped codlins and costards which lacked most of the qualities of the wild and abundant crab. Most of these Cumbrian place-names refer to the crab apple but it is possible that a few of the later 'orchard' names (-garth, -thwaite, -worth) may refer to the pippin which had become more widely known in England by the sixteenth century.

In Shakespeare's *Henry IV* Shallow invites Falstaff to his orchard where 'we will eat a last year's pippin of my own grafting, with a dish of caraways', and in *The Merry Wives of Windsor* Sir Hugh Evans dismisses the servant with the words 'I will make an end of my dinner, there's pippins and seese* to come.'

But the crab apple was still the popular festive fruit for the wassail bowl:

> *When roasted crabs hiss in the bowl*
> *Then nightly sings the staring owl,*
> *Tu-who;*
> *Tu-who, Tu-who, a merry note.*

For country folk, however, the crab apple still reigned supreme as it had done since apples were first brought to

---

* i.e. cheese

Britain by the Romans. The historical records of Anglo-Saxon and Norse writings show that by the tenth century the crab apple had established a firm place in the diet and mythology of Northern Europe. The Vikings who furnished the grave-ship of their Queen at Oseberg provided her with a bowl of small apples for use in the next world and as a symbol of fertility and eternal youth. The goddess, Idun, was the guardian of the apples which ensured that the gods would never grow old. As host to the sacred and magic mistletoe and as the source of the potent and magic verjuice (the fermented juice of the fruit), the apple was a tree anciently revered.

In Saxon times the apple began to be viewed in quite another light. A misunderstanding occurred in the translation of the Latin word *malum* which may mean both evil and apple, a mistake repeated in King Alfred's translation in the 890s of Pope Gregory's *Cura Pastoralis* (Pastoral Care). So the fruit of the unidentified tree of knowledge of good and evil which the Book of Genesis forbade Man to eat was henceforth believed to be the apple but was most probably a fig.

Fortunately, this biblical command, like so many others, was more honoured in the breach than the observance and although, as a raw fruit for eating the crab apples leave much to be desired, a few varieties could be more palatable as Bartholomeus Anglicus writing in 1470 indicated: 'Malus the Appyll tree... is gracious in syght and in taste and vertuous in medecyne... some beryth sourysh fruyte and harde, and some ryght soure and some ryght sweete, with a good savoure and mery.'

Cooked or stewed with honey, however, or pressed and fermented for a cider, they were much in demand. At Christmas festivities roasted crabs hissing in the wassail bowl were an essential part of the seasonal rituals long before the end of the Middle Ages. And it was not long before horticulturalists developed several of the sweet dessert apples with which we are familiar today.

In the literature, mythology and legend of all the cultures of Europe from the Greek and Roman to the Druidic, Celtic, Anglo-Saxon and Norse the apple is a constant feature. Even the legendary home of King Arthur, the Isle of Avalon, is said to be 'The Isle of Apples' and, in more modern times, the story is widely believed that it was an apple that revealed the mystery of the force of gravity to Sir Isaac Newton.

# ASH TREE

*Zwinger*

**Derivation:**    Old English: *æsc*
Old Norse: *eski/askr*

**Place-names:**  Asby, Ashness (1211), Askam, Aspatria (1160), Askew, Ashthwaite (1180), Askham (1232), Ashley-garth, Askrigg, Ashslack (1175), Askill, Esk Hause, Ashes, Eskin, Ashgill, Eskett (1230)

In the mythology of pre-Christian Scandinavia, and throughout much of northern Europe, the ash was a sacred tree. Yggdrasil, the World Tree, was said to be an ash with its roots in Hell and its canopy in Heaven. It was the seat of Odin, the ruler of the world of gods and men. It was also the home of the World Serpent, gnawing away at its roots and plotting the destruction of the whole of Earth and Heaven, but the tree is eternally renewed, providing life, shelter and nourishment to all living things for all time.

Even after the catastrophe of Ragnarok, the day of Doom and the cataclysmic end of the world, the ash tree survived, and, safe within its branches, were two beings, Askr and Embla, the Man and the Woman who were to re-people the new Earth which was created from the wreck of the old and who are today commemorated by a sculpture in Sweden and a wooden panel in Oslo City Hall. The parallel with the teachings of the Christian Church, with the Resurrection at the heart of its faith, is not difficult to see and it was inevitable that many of the ancient beliefs surrounding the ash tree should persist well into the Christian era.

Most persistent of all was the belief that an ash tree planted close to a dwelling was a sure protection against snakes. There was also a long-held belief that a sick child could be cured if it were passed through a cleft in an ash tree – the child healed as the tree healed. It was considered unlucky to break a branch from an ash tree; cows could be cured of 'the fairy sickness' by stroking them with an ash twig; ash 'keys' were worn as a protection against evil and witchcraft and as a cure for warts. Concoctions made from the leaves and the keys were prescribed for a variety of complaints, from the bite of an adder to flatulence and stitches in the side. Gilbert White's *Natural*

*History of Selbourne,* written little more than 200 years ago, indicates that much of all this was still part of country lore in his day and some of these beliefs are known to have survived even into modern times.

Even so, the countryman of almost any age was quite prepared to make use of the tough and pliable quality of timber from the ash tree. He knew that it matured more quickly than oak and that it could be used for more purposes than those of any other tree. Ash was a notable wood for farm and household tools and implements, for making carts and wagons, for long-lasting fences, for axe and pitchfork handles, and its aura of mystique made it the only wood appropriate for shepherd's crooks. As a fuel for burning on the hearth it surpassed all others.

There are few references to the ash as a culinary aid although John Evelyn notes that 'Ashen keys have the virtue of capers' and they were often used to flavour sauces.

In the twenty-first century the ash tree, steeped in mythology and folklore, sacred, venerated and cherished for centuries, is now under attack from the devastating disease of *chalara* or ash-dieback which threatens to destroy almost the entire species.

## ASPEN TREE

*Matthiolus*

**Derivation:**    Old English: *æspe*
                     Old Norse: *espi*

**Place-names:**    Aspland, Esps, Asphill (1603), Esp Ford, Esp-
barrow, Espland, Isthmus (Espness/Espenese 1210)

This delicate tree with its patterned silver bark and its quivering leaves, whispering in the breeze like gently falling rain, has an aesthetic beauty quite at variance with the legends which surround it. The unceasing trembling of the leaves, unique and unexplained before the age of scientific botany, was thought to be clear evidence of guilt or fear: 'to tremble like an aspen' passed into the language of everyday life.

Mistress Quickly in Shakespeare's *Henry IV* Part 2 expresses her detestation of 'swaggerers': 'Feel, masters, how I shake... yea, in very truth do I, an 'twere an aspen leaf: I cannot abide swaggerers.' The aspen's crime was not to have bowed, as every other tree did, when Christ made the journey to Calvary; its fear was of some eventual divine retribution; its guilt was to have provided the wood from which the Cross was made. From Russia to the Scottish Highlands and from Wales to Brittany the aspen was believed to carry this burden and so it is condemned to tremble for all eternity or until the day when all sins are forgiven. These gloomy superstitions have in modern times been replaced by an appreciation of the true beauty of the 'quaking aspen' as one of the most attractive of our native trees.

The unceasing delicate trembling of the aspen's leaves is sometimes believed to have a soothing, calming effect on the troubled human mind but on the other hand it was also thought that it could have such a disturbing influence that there were tales of people suddenly disappearing from beneath the tree, spirited away by the fairies. These mystical associations may have led to the strange custom of placing a crown of aspen leaves on the head of a body as it was interred in a burial mound in the belief that it would give the wearer the power to return from the underworld to be reborn.

A less than chivalrous but once popular contention was that

the leaves of the aspen and the tongues of gossiping women are made of the same material. Sir Thomas More, said to be a kind and tolerant family man, commented in a moment of desperation: 'These aspen leaves of theirs never leave wagging' and John Gerard described the quivering leaves as *langues des femmes*.

The wood of the aspen had few practical uses for the medieval farmer but as it is resistant to rot and does not easily warp it was occasionally used for roofing 'slates'. Beavers, too, seem to be aware of this and it is their preferred wood for their dams. The name itself is derived from the Greek word *aspis*, a shield, and both Greek and Celtic shields are believed to have been made of aspen wood because it is light and flexible.

Espness, the 'aspen headland' by Derwentwater, now called Isthmus, must have been a favoured spot in those distant days perhaps because of the beauty of the trees or, more probably, because of the aspen's many mystical associations.

# BEECH TREE

*Matthiolus*

**Derivation:** Old English: *bece / boc*
**Place-names:** Becha Potts, Beech Hill (seventeenth century), Bochestede (1294)

Julius Caesar stated that there were no beech trees in Britain and, while this was probably true in the first century, as far as

89

northern Britain was concerned, it was almost certainly incorrect for the chalk country of the south which is ideal for the beech. By early medieval times there is evidence that the easily worked fine-grained wood was preferred in those areas of the country for such common items as spoons, bowls, shovels, mallets and other household and farm implements and utensils. Beech, too, was the wood used to bind precious books such as the beautiful illuminated manuscripts found in monastic libraries. The Old English word for both beech and book is, in fact, the same word, *boc*.

Further north in the damper soils and limestone areas some beech trees had appeared by the early Middle Ages but the oak, birch, hazel and alder were much more common. In the south the water-resisting qualities of beech made it the favoured timber for mill sluices and for such enterprises as the piles for bridges and for weighty buildings such as Winchester Cathedral. In the north, the alder, with similar qualities, was much more plentiful and just as serviceable. We may reasonably assume that in the very few places where beech trees grew in Cumbria it was the wild boar and domestic pigs which appreciated them for the nourishing beech mast which carpeted the forest floor each autumn. Both the early place-names given are little known field names. The now lost Westmorland place-name 'Bochestede' dates from the thirteenth century but it is just one isolated name and it seems that the beech tree was not a feature of significance in the economy of the early medieval inhabitants of the north-west.

# BIRCH TREE

*Matthiolus*

**Derivation:**   Old English: *beorc*
Old Norse: *birki*
**Place-names:**   Barkrigg, Birk Bank, Birkbeck, Birkber, Birk
Crag, Birker (1279), Birker Beck (1205), Birkett (1210), Birk
Fell, Birk Hagg, Birk Howe, Birk Moss, Birk Rigg(s), Birks,

Birkshaw, Birkwray, Birker Fell, Birk Knott, Birk Dault, Birk Side, Birk Field, Birk Row, Birch Close

The silver birch is one of the most popular and easily recognised trees. Its silvery bark and graceful branches and, in early spring, the delicate tracery of its greening leaves shimmering against the sky delight the eye and lift the spirit. Dorothy Wordsworth noted in May 1802 that, 'The Birch Tree is all over green in small leaf, more light and elegant then when it is full out' whereas in winter the birch is 'beautiful red brown and glittering.'

It is easy to understand why so decorative a tree came to be endowed with certain qualities of purity and innocence. The Celts and the Germanic tribes of Europe revered the birch as a Holy Tree with the power to drive out evil spirits. A ritual was evolved in which a 'beater' made of birch twigs drove out from house, barn and byre, all that was evil or malevolent as the old year gave way to the new. The adaptation of this idea was to be lamented by many later generations of delinquent schoolboys and errant wives. Whether these beliefs and disciplinary customs associated with the birch tree prevailed in the Norsemen's Cumbria we do not know – social customs in pre-Christian Scandinavia would suggest not – but we do know that a form of Tree Worship was common and that there was an abundant supply of birch trees.

The fine-grained wood was used extensively for household utensils as it is easily carved or turned; the twigs made excellent besoms for sweeping; the bark, in sizeable rolls or strips, made a good waterproof 'felt' to place under the roof turf or thatch.

We may be reasonably sure, too, that some of the hard-drinking men-folk were well acquainted with the birch tree's potential as a source of a powerful alcoholic spirit. From an early spring incision in a mature tree some fifteen gallons or so (about 66 litres) of thin, sugary sap could be drawn off and after suitable fermentation would help many a long, dark evening to pass more quickly. There is some evidence that the bark of the birch soaked in water was used as a form of plaster-cast for broken limbs as it acquired a firm stiffness as it dried. More prosaically, birch wood makes excellent firewood and we may be reasonably sure that this was one of its more important uses in the medieval centuries.

The Industrial Revolution had a devastating effect on the birch woods of Cumbria. Whole forests were felled and we may count ourselves fortunate that so much has recovered. Some idea of the huge demand made on the birch tree at this time can be gained from an article in the *Penny Magazine* for 1843:*

> Big trunks can be turned, used for furniture, mills, carts, ploughs, gates and fences. Small branches make hoops, besoms, ties for faggots, baskets, wicker hurdles and all turnery. The bark is much employed for tanning leather and also yields a yellow-brown dye. Its charcoal burns a long time and is much in demand for making gunpowder and crayons. The ashes are rich in potash.

The profusion of birch tree place-names is a clear indicator of the extent of the birch woods and of the importance attached to them. Some have been lost, including Birchenesfelde which we now know as the Buttermere fell 'Robinson'.

---

* Quoted in *Industrial Archaeology of the Lake Counties* by J. D. Marshall and M. Davies-Shiel, 1969, p163.

*Gerard*

## BIRDCHERRY OR HECKBERRY

Derivation:     Old Norse: *heggr*
Place-names:    Egholme (1279), Hegdale (1201), Heggerscale (1380)

The birdcherry is mainly a northern tree and from Lancashire and Yorkshire northwards it owes its local names to the Scandinavian peoples who settled there: Hawkberry, Hagberry, Hegberry, Hackberry, and in Cumbria, Heckberry. Dorothy Wordsworth describes a walk in Easedale from where she 'brought home heckberry blossom'. The long, white flower spikes of the heckberry wafting an almond fragrance on the breeze are one of the delights of springtime but otherwise the tree has little to commend it.

The timber is of no practical value but its unpleasant smell may have led to the belief held by some that, placed by the door of the house, it would provide protection against the plague. The berries are exceedingly bitter to taste and, as the name implies, only palatable to birds. Only the bark was considered to have a certain dubious virtue and in some parts of the country an infusion made from the bark was believed to be useful for stomach ailments, but there were many other, less bitter and more easily acquired, herbal remedies for such complaints. There was also a belief that a piece of bark placed in the drinking water was a protection against disease but it is difficult to imagine that the inhabitants of Cumbria felt that the pure water in their fast-flowing becks presented any danger. A few may have spiced their ale with the bitter black berry but this must surely have been an acquired taste.

Hegdale, Heggerscale and Egholme may have been noteworthy places quite simply from the heckberry's lavish display of blossom each spring and not from any particular practical value of the tree itself.

## BLACKTHORN OR SLOE

*Matthiolus*

**Derivation:**  Old English: *slá*
**Place-names:**  Slaythorns, Sleethorn, Slealand, Sloebank, Sleathwaite, Sleagill, Sloethorn, Sleastonhowe

'I walked to Rydale after tea… The Sloe-thorn beautiful in the hedges and in the wild spots higher up among the hawthorns.' Thus Dorothy Wordsworth in the spring of 1802 records her pleasure in seeing one of the delights of early spring. The drifts of the blackthorn's delicate white blossom covering the hedgerows have brought pleasure and hope, after a hard winter, to many generations of country folk. Experience taught, however, that the flowering of the blackthorn did not mean that winter was finally over. 'Beware the blackthorn winter' was one of the sounder pieces of weather-lore: so long as the blackthorn blossomed a sharp late frost could strike at any time.

In the earliest Anglo-Saxon charters, long before maps were drawn or hedgerows were planted, the boundaries of parishes and landholdings were often marked by prominent bushes of blackthorn or hawthorn, and many of them became well-known landmarks.

Throughout the medieval centuries the blue-black berries of the blackthorn (or sloe) were gathered soon after the first frosts, to be chopped and grated in honey to make their sharp, dry tartness more palatable. For, as Nicholas Culpeper put it, the fruit of the sloe is of 'a rough sour austere taste'. Today it is rarely eaten in this way as we have become accustomed to sweeter fruits, but, mixed with apples, it produces a jelly appreciated by connoisseurs. Many people today will be more familiar with a type of liqueur known as sloe gin, not even a close relative of genuine gin but a pleasant enough drink and at one time a popular home-made product.

The sloe is unlikely to be found in any modern pharmacy although Culpeper, Gerard and other herbalists praise its medical properties. 'The juice of Sloes do stop the belly' according to Gerard who adds that they 'may well be used instead of Acatia which is very hard to be gotten and of a deere price; albeit our plums of this countrie are equall to it in vertues.'

Culpeper tells of another, little known practical use made of the sloe in his day: It is the juice of this berry that makes the famous marking ink to write upon linen, an ingenious way to solve an identification problem when the only other ink available was made from soot, gum and water.

Blackthorn was the wood traditionally used for walking sticks, notably the Irish shillelagh and for the wands of wizards and witches. Harry Potter had to use a blackthorn wand after Hermione broke his superior holly wand.

## ELDER TREE OR BURTREE

*Matthiolus*

**Derivation:**    Old English: *ellærn* (mainly in southern counties)
Northern dialect 'Burtree' or 'Bortree' from Old Norse: *bora*

**Place-names:** Burtree (13th century), Bortree, Burtree Bank, Burtree Gill, Burtree Hill, Burtree Hole, Burtree Scar, Burtergill, Burtresdale (1199)

Pungent masses of creamy-white springtime flowers and drooping clusters of purple-black autumn berries make the elder among the more easily recognisable trees in the English countryside. In northern counties it was universally known as the Burtree or Bortree, a reference to the ease with which the pith can be removed from the stem to leave a clean hollow 'bore'. Pliny, two thousand years ago, noted this as one of the many attributes of this tree, adding that in this way it could be used to make whistles and musical pipes. Nicholas Culpeper, sixteen centuries later, confirmed that this age-old custom was still well-known and had been adapted to other uses by the ingenuity of youth: 'Every boy that plays with a pop-gun' he tells us, 'will not mistake another tree for the elder.'

Whatever the virtues of the elder in the manufacture of wind-instruments and pea-shooters, it was not for this but for its many qualities as a medicine that it acquired a reputation and respect exceeding that of any other plant or tree. For some two thousand years it was the indispensable medicine of country folk throughout Europe. It was cherished by Hippocrates, the Father of Medicine, in Ancient Greece; it was also highly regarded, as Pliny tells us, in Ancient Rome; it figures prominently in the Anglo-Saxon, Norse and Medieval Herbals; by the time of Shakespeare it had been raised to the level of the greatest of them all:

*What says my Aesculapius? My Galen? my heart of elder?*

In 1644 the elder merited a special treatise – *The Anatomie of the Elder* – which listed over seventy ailments or diseases for which some part of the tree, suitably prepared, would provide a remedy. Every part of the elder had some medicinal virtue, we are told; almost every bodily illness could be cured by it.

A few years later John Evelyn was equally convinced of this: 'If the medicinal properties (of the elder) were fully known I cannot tell what our countryman could ail for which he might not fetch a remedy from every hedge.' Flowers, roots, leaves, bark and berries were all beneficial, so much so that the great Dutch physician, Herman Boerhaave, who set new and more scientific standards of medical research in the late eighteenth century, raised his hat every time he passed by an elder tree in acknowledgement of its unique services to humanity.

A decoction prepared from the roots acted as a somewhat violent purgative when a thorough cleansing was considered desirable; it could also be used, diluted, as a gentler remedy for dropsy. The bark also was recommended by Hippocrates for the same purposes but the diluted nature of the prescribed dose (one ounce in a pint of water taken in small quantities), indicates its potency. The leaves made an even more nauseous purgative and were mainly used to make 'Green Elder Ointment', a widely known remedy for piles, bruises, sprains, chilblains and wounds of all kinds. Bruised leaves or the juice of the leaves were used in house and byre to drive away flies and other insects and also in stores and granaries to keep out the mice. It was this repellent effect which led Shakespeare to refer to the stinking elder.

John Evelyn believed that a gruel made from the young buds 'effected wonders in a fever', and Nicholas Culpeper maintained that the young shoots 'boiled like asparagus, and the young leaves and stalks boiled in fat broth, doth mightily carry forth phlegm and choler.'

Roots, bark and leaves were clearly versatile in their medical applications but one suspects that, in certain cases, the effects of too strong a dose could be stunning. In fact, the berries and most other parts of the elder are distinctly toxic before they are boiled and one must suspect that some of our distant ancestors only discovered this by trial and error.

The flowers are much more benign. Evelyn, again, has much to say in favour of the elder flower. He relates that a small ale in which elder flowers have been infused was so much in demand in seventeenth century London that it 'is to be had in most of the eating houses', and elder flower vinegar obviously added relish to his salads. Every book of herbal medicine has extolled the merits of elder flower ointment and an infusion of the flowers is still recommended to treat sore throats, fevers and colds; and it is not so many years since Elder Flower Water, to freshen the eyes and to keep the skin clear and soft, was part of every lady's toilet equipment. As the main ingredient of a pleasant cooling drink the elder flower is still often gathered in June and, in former days, enough flowers were collected to pickle in salt to ensure a supply for the rest of the year.

Almost all the ailments which could be treated with the leaves and flowers would also respond to elderberry wine, particularly as a hot nightcap. The adulteration of port wine with a proportion of elderberry wine was a dubious commercial activity in the eighteenth century and led to the belief (not entirely without foundation) that this combination could be an effective treatment for sciatica and neuralgia. The extensive claims made for the elder by a long succession of herbalists may all be given a certain modest degree of credence but even they did not go quite so far as John Evelyn who, in an excess of enthusiasm, pronounced that 'indeed, this is a catholicum against all infirmities whatever.'

Medicine and magic are never very far apart and it is not at all surprising that the elder, above all, should over so many generations acquire its share of folklore and superstition. As with other 'magic' trees, the elder was believed to have the power to ward off bad luck, evil spirits and witchcraft. Even today there are hedge-cutters who always leave an elder uncut; and those out gathering firewood who will never take an elder branch for burning. In more religious times crucifixes made of elder were hung over the doors and windows of stables and byres and even cottages, no doubt to fortify the mystic forces in the sprigs of rowan, ash and hazel already in place.

In Norse mythology the elder tree was the home of Hyldemor, the Elder Mother, who would haunt all those who cut her tree without her permission. This must have been readily granted for elder-wood was much sought after to make bowls, spoons, pegs, skewers, combs, children's toys and many farm and household implements and utensils. In the great enclosing of land in the seventeenth and eighteenth centuries elder was often recommended as the best tree to cut for fences as the posts would last for so many years.

On the other hand, Hyldemor wielded magical powers which would bring misfortune and bad luck to those who did not treat her with respect. A wise person would always ask Hyldemor's permission before cutting an elder tree for any purpose. Thus, for example, it was prudent to consult Hyldemor if you wished to make a baby's cot or cradle from elder wood – and it was probably sensible not even to consider the idea.

The number of 'Burtree' names in Cumbria would seem to indicate that, for medical, household and superstitious reasons, the bortree or elder was as highly esteemed in the north as the Ellan-tree was in the south.

# ELM TREE

*Gerard*

**Derivation:**     Old English: *elm, wic*
Old Norse: *almr*

**Place-names:**  Alme Bank, Elm Bank (c.1200), Alm Gill, Alm Slack, Elme Tree, Elm How

The English elm which not long ago succumbed to the ravages of Dutch Elm disease was a comparative rarity in the English countryside until the landscape artists of the eighteenth century made it a feature of the fine parklands of the stately homes then being built in all parts of the country.

The much hardier Wych Elm is more commonly found in the north of the country and is almost certainly the tree referred to in these Cumbrian place-names, particularly in the older names. Wych is an Old English word – *wic* – meaning pliable, and it was this quality, together with its durability and toughness, which made wych-elm timber so attractive in an age when wood was so much in demand. It is almost impossible to split and so was ideal for the hubs of wooden wheels, the shafts of carts and the floors of threshing barns. It was also used for milking stools, mangers and the stalls of the cow byres. An elm-wood bread trough was a valued possession of every housewife.

Elm is notably durable in damp conditions and was later put to use for water pipes, drain pipes, mill sluices and the keels of boats. It was also the wood traditionally used for coffins.

The pliability of elm proved immensely appealing to the medieval bowyer. Gerard of Wales stated that the bows used by the Welsh bowmen were 'made neither of horn, ash nor yew but elm.' The long English long bow was normally made of yew. It was the long bow which provided both the England and the Welsh with the weapon which gave them dominance in warfare and in hunting for so many years.

The medieval Herbals have little to say about the medicinal uses of the elm but Nicholas Culpeper makes a number of somewhat bizarre claims in this respect. Among other unlikely prescriptions he states that 'a decoction of the leaves, bark or root heals broken bones', adding that 'the water from the bladders on the leaves if put in a glass and set in the ground or else in dung for twenty-five days… is a singular balm for green wounds; and if the roots of the elm are boiled in water for a long time, the fat arising on the top thereof used as an anointment will cure baldness.' No doubt there were desperate folk who had tried all these.

Equally desperate must have been the country folk of Norway who in the serious famine of 1812 resorted to eating strips of boiled elm bark to assuage their acute hunger. It is not impossible that their Norse ancestors in Cumbria may also have been reduced to such extreme measures when a failed harvest was followed by a severe winter but they would have been well aware of the many practical uses of the elm tree.

# HAWTHORN

*Matthiolus*

**Derivation:**   Old English: *haga-thorn*
Old Norse: *hagi-thorn*

Place-names:    Thornflatt, Thornship (1226), Thornholme, Thornbank, Thornbarrow, Thorney, Thorny, Thorny Scale, Thorny Gill, Thorneygale, Thrimby, Thornyfields, Ullathorns, Thornhow, Thornthwaite, Thorny Plats, Thornwray, Thorny Slack, Thorny Moor

The large number of northern place-names referring to the hawthorn – some of them first recorded in the thirteenth century – suggests that this was a fairly common tree from an early date. It appears in many Saxon charters as a boundary marker for parishes and manorial estates. Later, at the time of the great enclosures in the eighteenth and nineteenth centuries, the southern counties, (and also some parts of lowland Cumbria) acquired many miles of hawthorn hedgerows, as required by the law, to define the limits of the new fields.

This was the time when upland Cumbria was covered with its network of drystone walls built for the same purpose. Thus, in much of Cumbria, the hawthorn is found not in extensive hedgerows but in clumps over the fell-sides or often in isolated specimens. Here, too, however, the strange connection of the thorn with the law is continued: a fifteenth century record refers to a thorn tree near Hesket-in-the-Forest which marked the place where the tenants of Inglewood Forest held their Forest Court to determine any legal problems. One of Cumbria's most famous landmarks was the Shap Thorn originally planted on a prominent hilltop to guide travellers in this sometimes-bleak area. (It has now been replaced by a clump of ash trees).

Much of the superstition and folklore attached to the hawthorn came to Britain from France rather than from Scandinavia where similar customs were associated with the rowan tree, as hallowed a tree in Norse mythology as the hawthorn is in the Latin countries where it was a powerful symbol of sex and fertility. In May the hawthorn is a glorious picture of white blossom with a heavy scent reminding one, according to age, mood and taste, of sex or putrefaction. By association the latter led to a belief that the hawthorn had magical properties which would protect against the plague. Similarly the hawthorn's association with sex and fertility featured lustily in the pagan customs surrounding the may-time Beltane festivities. Folklorists believe that the original maypole – a phallic symbol in itself – may have been a hawthorn tree and that the ribbons wound round the maypole represent the union of male and female.

Rosalind in Shakespeare's *As You Like It* refers to an old and more romantic custom associated with lovesick swains and not unknown even today: 'There is a man haunts the forest, that abuses our plants with carving 'Rosalind' on their barks, hangs odes upon hawthorns and elegies on brambles.'

On May-day (12 May in England until 1752) sprays of may-blossom were at the heart of much lively and licentious ceremony and rituals dating from pagan times which the Christian Church attempted to suppress or to adapt to its own uses. To the young and nubile, even in priest-dominated medieval times, may-blossom time was a time for celebration and revelry; to their parents, the value of the hawthorn was as a source of tough hard wood for tools and implements and, especially for the housewife, its wood made the hottest fire for the bread oven. For those who had suffered from contact with the formidable thorny defences of the hawthorn an ointment concocted from the leaves and berries made an effective poultice for drawing out thorns and splinters. But perhaps the hawthorn's most important value to medicine is its use –

known to the Druids and to modern medicine alike – in the treatment of high blood pressure and many cardiac problems such as angina and irregular heart rhythm.

The hawthorn features in the heraldry of the Tudors but there is no historical evidence to support the long-held belief that the Crown of England was discovered hanging on a hawthorn bush on the Bosworth battle field. It was not until 1898 that this became part of the Bosworth story when the eminent historian, James Gairdner, presented it, without any supporting evidence, as historical fact. Shakespeare would surely have made use of its dramatic potential if such an historically bizarre incident had actually occurred.

The large number and great variety of hawthorn place-names in Cumbria leave no doubt that the heavy scent of the white swathes of may blossom in spring and the display of scarlet berries in autumn have been a feature of the Cumbrian landscape for a very long time.

# HAZEL TREE

*Matthiolus*

**Derivation:**   Old English: *hæsel*
Old Norse: *hesli*
**Place-names:** Helsington (1086), Haslehevet (1200), Hazelseat, Hazelslack (1254), Hazelspring (1272), Hazelbank, Hazelsike (1331), Hazelrigg (1285), Hesley, Hazelber,

112

Hazelgill, Hazelhurst, Hazelshaw, Hazelbeck, Hazelholme

The hazel, like the rowan and the hawthorn, was a sacred tree, associated with ancient pagan rituals. Regarded as the 'Tree of Knowledge' it was held in particular esteem and was believed to confer special powers of wisdom and protection. Egil's Saga relates that the judges and law-makers in Norse communities deliberated inside a fence of hazel staves known as the 'vebond' or 'sanctuary boundary', a custom described by W. G. Collingwood in his novel *Thorstein of the Mere* in reference to the Norse Althing (council) held at Legburthwaite in the Vale of St John.

Chiefs and lawgivers were often buried with a cross in the form of two hazel wands as the Christian Church attempted to find a place for the 'magic hazel' in its Christianisation of pagan customs. Kentigern, the Cumbrian missionary, is said to have blown upon a green hazel twig to create a flame to light the altar candles, and the church even dedicated the hazel to St. Philibert and changed its name to the filbert, but these subterfuges made little progress in diminishing popular belief in the magic powers of the hazel, particularly in matters concerning sex, fertility and witchcraft. Indeed, with the re-discovery of the water divining powers of the hazel twig in Tudor days the old customs gained a new lease of life and remained part of the Maytime festivities until modern times.

On a more practical level the hazel had always been valued for making fences and hurdles, for thatching-spars, for the wattle of 'wattle and daub' in house construction, for baskets, swills and cradles, and, of course, for its abundant supply of nourishing nuts.

The Industrial Revolution created a demand for flexible

hazel rods which far exceeded the natural supply and so the system of coppicing began. Every seven years the hazel was cut for use in the manufacture of swills and corves and for making hoops for barrels. The gunpowder mill at Gatebeck near Kendal required over 4,000 barrels each year between 1866 and 1914; mines throughout Cumbria used hazel corves to haul coal to the surface for more than 200 years. If one adds that 40 tons of hazel nuts were shipped from Broughton-in-Furness as part of one year's exports during the nineteenth century, it is not surprising that many of the place-names are first recorded at this time, although some have a much earlier medieval origin.

One aspect of the ancient veneration for the hazel is depicted by the Brothers Grimm in their tale *The Hazel Branch*. Mary, the Mother of Jesus Christ, was out in the woods one day gathering strawberries when she was attacked by an adder. She took refuge behind a hazel bush and its magical powers warded off the adder which crept away leaving her unharmed. And so, since then, according to the tale, a green hazel branch has been the safest protection against adders, snakes, and everything that creeps on the earth.

# HOLLY TREE

*Gerard*

**Derivation:**     Old English: *holegn*
Old Norse: *bein-vithr*

*Place-names:* Hollin Bank, Hollin Brow, Hollinghead, Hollingwell, Hollinhow, Hollinrigg, Hollinroot, Hollinshaw, Hollinslack, Hollinthwaite, Hollins, Hollens, Baynwythrig (1220), Bannerdale (1202), Bennet Head (1285)

Writing in the first century of the Christian era Pliny the Elder, in his Natural History, gave classical authority to what was already an ancient and widely held belief that the holly tree had a special power against evil spirits, demons, goblins, and supernatural spells of all kinds. The dwellings of both man and beast were protected from lightning and poltergeists and the wiles of witchcraft if holly trees were planted nearby.

The magic potency of the bright red berries was believed to be particularly effective in the fearful dead of the year. The Christian Church had little difficulty in adapting this pagan veneration for the holly to its own purposes. The symbols were numerous: the white flowers represented the purity of Christ's Nativity, the red berries Christ's blood shed on the Cross, the prickles Christ's Crown of Thorns, the evergreen leaves were a symbol of Eternal Life, and the bark, 'bitter as any gall', was a remembrance of the suffering of the Crucifixion, with the clean, white wood a fitting material for Christ's Cross itself.

It was not long before new Christian names began to appear for the holly: the Holy Tree, the Christmas Tree, Christ's Thorn and Prickly Christmas. The pagan custom of 'protecting' the house against evil spirits with holly branches and their powerful red berries was quickly adopted to the Christmas celebration of Christ's Nativity. From Christmas Eve until Candlemas Eve the holly became a traditional part of the mid-Winter festival. Christianity had confirmed that the holly

116

was, indeed, a very special tree:

> *Of all the trees that are in the wood*
> *The Holly bears the Crown*

Shakespeare's well-known song in *As You Like It* would seem to confirm this and with it the place of the holly in the Christmas festivities:

> *Sing, Heigh-ho! unto the green holly!*
> *This life is most jolly.*

The inhabitants of Brough in Westmorland obviously concurred with this, for it was their custom on Twelfth Night to hold a torchlight procession with torches tied to holly branches as part of their mid-winter celebrations.

The predominance of Old English place-names recording the holly may well reflect the early Christianisation of the Angles and Saxons in Britain. The holly was less familiar to the Norsemen who appeared on the scene several centuries later, and the Norse name, *beinvithr,* 'the bone-wood tree', soon disappeared even in Norway where it was replaced by Kristtorn, 'Christ's thorn'. The three Cumbrian place-names for the holly which are of Norse origin all have a thirteenth century date. These may be merely the first written records of much older names rather than evidence of contemporary survival of the Norse language.

Thomas West in his *Antiquities of Furness* relates that it was the custom in Furness for shepherds to feed their flocks on the tender sprouts of the holly and holly trees were carefully preserved for that purpose so that there were large tracts of common being so covered with these trees as to have the appearance of a forest of hollies. The mutton from sheep fed on this diet was said to have a particularly attractive flavour.

The ancient belief in the magic qualities of the holly is reflected in J. K. Rowling's books on the adventures of the boy wizard, Harry Potter, where we learn that his wand was made of holly – and far superior to the blackthorn substitute he had to use when Hermione broke his precious holly wand.

# LIME TREE

*Gerard*

**Derivation:** Old English: *lind*
Old Norse: *lind*

**Place-names:** Lindale (1191), Lindal, Linbeck (1200), Lincrag, Lind End, Lindemire (1241), Lindeslac (1200), Linbeck, Lindeth (1262), Lindreth (1616), Lyndhowe (1358)

The small-leaved lime is native to Britain and the early dates of most of the Cumbrian place-names suggest that this is the tree which gave its name to Lindeth, Lindale, Linbeck and all the other places where this most decorative tree grew. The larger limes were not introduced until the seventeenth or eighteenth centuries. All the names occur in the southern parts of Cumbria – Furness, South Westmorland and the lower reaches of the Esk – for this is the northern limit of the small-leaved lime in Britain. Cumbria is home to one of the oldest small-leaved lime trees in the country. In the Forestry Commission's Rainsborrow Woods in the Duddon Valley is a tree believed to be more than 500 years old.

John Gerard described the wood of the lime as very soft and gentle in the cutting and handling and this was most wonderfully demonstrated in the remarkable and delicate woodcarvings of Grinling Gibbons, notably at Hampton Court and Windsor Castle. Several hundred years before Gibbons the monks of Furness Abbey were displaying considerable skill and business acumen in using lime-wood to fashion cups, bowls, dishes, spoons and other useful utensils. The Dissolution Commission in 1539 made a note of this and also reported that they made a coarse matting from the 'bast' or fibre of the inner bark, a material also used to make lightweight ropes or small 'ties' for young trees.

The lime – or linden – tree has important medical virtues which have been known for over a thousand years. The flowers are used to produce Linden tea a pleasantly flavoured drink which has a well-known reputation as a soothing agent to relieve anxiety, to calm the nerves, to aid digestion, to offer a remedy for sleeplessness and to relieve some of the problems of a sore throat and the common cold. The delicious fragrance of the flowers attracts honey bees to the nectar which produce a highly prized white honey.

A further reflection on the high regard in which the lime tree has been held for so many years appears in the moving story of *The Man Who Planted Trees* by Jean Giono. Over many years a shepherd in Provence created a large new forest on barren moorland where once a busy farming community lived. A visitor returning to a revived village and landscape discovers that by the village fountain a linden tree had been planted which was already in full leaf, an incontestable symbol of resurrection.

The lime tree was one of the 'holy trees' often planted in villages in earlier times to protect the inhabitants against the forces of evil.

# OAK TREE

*Matthiolus*

**Derivation:**     Old English: *ac*
Old Norse: *eik*
Brittonic: *derw*

**Place-names:** Aikbank (1292), Aikhead (1270), Aigill Sike (1220), Aikshaw (1292), Aikton (1200), Aikbeck, Aiken Kott, Aikrigg, Oakbank (several), Oaks, Oakhowe, Oakshaw, Oak Hill, Oakhurst, Oakland, Derwent River, Derwent Fells, Derwent Isle, Derwentwater

*Quercus robur*, the hard, strong oak, has been to every age the symbol of endurance, reliability and strength of character. Its wood is so hard that it was almost impossible to cut before the invention of the iron axe and, partly because of this, the oak was held sacred and worshipped by the Greeks and Romans and by the tribes of Gaul and ancient Britain. The Druids held it in particular veneration and gathered the 'baleful' mistletoe* from its boughs for their religious rites.

The Norsemen valued the oak for a more practical purpose: their famous longboats were largely built of it, the keel itself being a single oak timber over seventeen metres long and about forty centimetres deep. By the time of the English and Norse settlements in Cumbria iron tools were well advanced and the attack on the great oak forests which covered the valley slopes and the fell-sides up to more than 1,000 feet began in

---

\* Shakespeare's 'baleful' reference reflects the Christian rejection of the mistletoe because of its association with Druidical sacrificial rituals and with sex and fertility. The pagan Norsemen also rejected the mistletoe because in their mythology it was with an arrow made from mistletoe that Balder, the son of Odin and Frigg, and their adored god of light, was killed. Eventually however, this changed and, probably in tribute to Balder's gentle nature, the mistletoe became, in the Norse culture, a symbol of peace and reconciliation, and by the eighteenth century in England this had developed into the custom of 'kissing under the mistletoe'.

earnest. For the next 900 years oaks were felled for a multitude of purposes – for the beams and frames of houses, barns and byres, for the axles of water-wheels, for ships and for household furniture. To all these the Tudor and Stuart ages added bigger and more splendid houses with fine oak panelling, and bigger and more splendid ships for the navies of Henry VIII, Elizabeth I and Charles II. More than 3,000 oak trees had to be felled for the construction of every battleship and the Tudor and Stuart navy comprised over 40 ships, all needing constant replacement. In 1578 a Notice of Sale offered '22,674 oke trees' in Ennerdale at sixpence a tree, a foretaste of the great felling of the oak woods which was about to start.

The annual crop of acorns from oak trees provides food for a number of birds and animals. Jays and squirrels not only consume them but hide them in caches to be recovered later and they appear to be endowed – some more so than others – with memory 'maps' to identify the locations where the hoards were buried. Woodpeckers also feed on acorns and deer seek them out as a high proportion of their autumn diet. Pigs also are able to tolerate the tannins in acorns and in medieval times the right to allow pigs into the forests to feed on this annual bonanza – known as pannage – was written into feudal terms of tenancy. In times of severe food shortage acorns have also been adapted for human consumption. Thoroughly soaked in water to leach out the tannins they have been ground to make bread flour and as a substitute for coffee, but neither was a family favourite.

The oak also produces an alien peculiarity commonly known as the oak-apple. This is a gall which develops after a species of wasp invades the leaf buds. Early human civilisations discovered that these galls could be the source of a black ink and from at least the age of the Roman Empire until the nineteenth century this was the main source of ink in the regions where oak trees grow.

As the Industrial Revolution developed in the eighteenth century, the oak forests of Lakeland provided resources for charcoal making for the iron and gunpowder and other industries. The best 200 year old oaks were felled for gunpowder barrels, young oaks for many thousands of spelks and swills, and a constant supply of bark for the tanning industry. Now the sessile oak – *Quercus petraea* – was in demand also, and already by the early nineteenth century, as Wordsworth's observed, there were few ancient woods left in this part of England.

The future for the oak looked bleak indeed, for even the young growths of natural regeneration were often eaten off by grazing sheep. Only the eleventh hour work of conservation makes it possible for us to enjoy the pleasures of the lovely oak woods of Borrowdale, Buttermere, Langdale, Eskdale and High Furness. In 1895 W. G. Collingwood was still able to record the oak tree dressing ceremony which took place on Maundy Thursday at Satterthwaite fountain, an interesting survival from an ancient pagan ritual, but the stark fact remains that the once sacred oak almost disappeared and became no more than a place-name.

The pagan worship of the oak tree was certainly part of the Celtic world and featured prominently in the rituals of the Druids. The very name 'Druid' is said to be derived from the Celtic *derw-weyd* meaning 'oak tree prophet'. The Christian church failed to eradicate the cult of the oak tree and, in various forms, it survived into modern times.

The celebration of the Restoration of the Monarchy on 29

May 1660, was known as Oak-apple day when sprigs of oak leaves were worn well into the twentieth century. Many oak tree which have survived to a great age or have some historical association are still objects of veneration, among them the Major Oak in Sherwood Forest, for ever linked to Robin Hood, the Boscobel Oak where historical legend relates that Charles II hid to escape from his Cromwellian pursuers, Birnham Oak in Scotland famously referred to in Shakespeare's *Macbeth*, the more than 1,000 years old Bowthorpe Oak in Lincolnshire, believed to be the oldest tree in England – all visited by thousands every year.

The oak tree has appeared from time to time on English postage stamps and on the national coinage; it is also the symbol used by the National Trust and the Woodland Trust.

# OSIER

*Gerard*

**Derivation:**   Old English: *withig, wiker,*
*tænel* (an osier basket)
Old Norse: *vithir, vikir*

**Place-names:** Witherslack, Wickerslack, Taernside(1220), Wickerfield, Wickerthwaite, Whitbeck, Whyber, Wyber, Wythop, Wythe Sike, Wythburn, Wythmoor, Wythestob, Wythes, Wiseslack, Wisetubland, Widewath, Widdygill

The osier is a plant of the willow family and flourishes in swampy hollows, marshes and damp places often near lakes, rivers and becks. The long stems or withies – the 'green willow' of so many songs and ballads – have been harvested throughout the ages to make every kind of household and industrial basket and container as well as cradles, chairs, fish traps and ropes. It is often known as 'the basket willow'.

One of the discoveries at the Viking settlement of Jorvik was a section of a house wall made of withies woven into stakes, a kind of wattle which became common in medieval house construction and was also used for hurdles and fences until very recent times. In the wetlands of Somerset the withy is still harvested for a variety of purposes but extensive drainage for agriculture has made it much less common in Cumbria than formerly. The large number of place-names referring to the withy is a reminder that the osier once grew here in abundance and was much in demand.

In folklore the willow has, for centuries, been associated with sorrow and grief. This seems to have its origin in the Book of Psalms where it is written:

*By the rivers of Babylon, there we sat down, yea, we wept, when we remembered Zion. We hanged our harps upon the willows in the midst thereof.*

Willow branches were taken into churches on such occasions

128

as Good Friday or funerals. In the course of time the willow also came to symbolise the grief of unrequited or lost love and often foretold death or tragedy. This was the inspiration for the famous Elizabethan ballad which Shakespeare adapted for Desdemona's song in *Othello*:

> *The fresh stream ran by her, and murmur'd her moans;*
> *Sing willow, willow, willow;*
> *Her salt tears fell from her and soften'd the stones;*
> *Sing willow, willow, willow.*

Shakespeare does not include the final lines of the ballad but the Elizabethan audience would know that they foretold the death of Desdemona:

> *Take this for my farewell and latest adieu;*
> *Write this on my tomb, that in love I was true.*

These were the sentiments of the age of the Renaissance. In earlier times when the lives of so many were, as Thomas Hobbes described them, 'solitary, poor, nasty, brutish and short' the various practical uses of the willow were of far greater interest.

# PEAR TREE

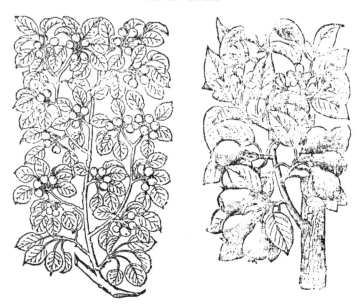

*Gerard – left the wild pear and right the great choke pear.*

**Derivation:**   Old English: *pere*
**Place-names:**   Peertregarth, Pear Crofts, Peartree Holm

The pear tree is not a native of Britain and may have been introduced by the Romans but this is far from certain. However,
pear trees are mentioned in Domesday Book (1086) as boundary markers so they may have been known in England before
the Conquest but it is more likely that they were brought by
the Normans from the Mediterranean lands.

The Norman nobility were particularly fond of perry and were less addicted than the English and the Norsemen to mead and ale. Pears remained a luxury imported from France throughout the Middle Ages: King Henry III (1216-72) was presented with a gift of Normandy pears by the City of London; this was not yet a fruit enjoyed by most of his subjects in England.

It would also take several centuries for this new fruit to spread to the remoter northern parts of England and all the place-names are first recorded in the seventeenth century, by which time Nicholas Culpeper could write that pear trees are so well-known that they need no description, and John Gerard had already pronounced perry to be 'a wholesome drink which comforteth and warmeth the stomach and causeth good digestion'.

Both Culpeper and Gerard refer to several types of pear, the wild and the cultivated, the sweet and the 'harsh'. Gerard also illustrates 'the great Choke peare', while the poet, Sir John Suckling, wrote of the red-striped Catherine pear, and the famous sixteenth century nursery rhyme has 'a golden pear' growing doubtfully on the little nut tree.

The Elizabethans had become familiar with a famous variety of pear: in *The Winter's Tale* the 'clown' charged with the preparations for a feast specifies that, among many other exotic items, he must have saffron to colour the warden pies, a reference to a famous cooking pear created by the monks of Warden Abbey in the fourteenth century and to a dish which was one of the highlights of medieval cuisine. It was not until 1780 that the mysterious Christmas song *The Twelve Days of Christmas* with its partridge in a pear tree first appeared in England.

Which of the varieties of pear tree were to be found in the Cumbrian orchards it is not possible to determine but clearly there were northern folk eager to sample the fruit and drink which had become so popular in the south. All three place-names indicate an orchard or enclosed area. The large, luscious pears we buy today did not appear until the nineteenth century.

# WILD PLUM OR BULLACE

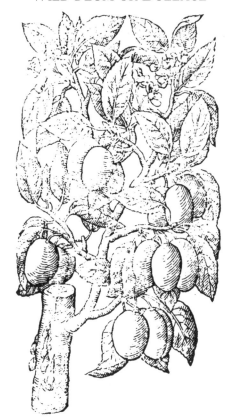

*Gerard*

Derivation:  Old English: *plume*
Old Norse: *ploma*
Middle English: *bolace*
Old French: *beloce* (sloe)

133

**Place-names:** Plumbland (1150), Bullace Croft, Plumpton Wall (1578), Plumpton (1212), Plomtreelands (1686), Plumgarth (1615)

Modern dessert plums are the fruits of horticultural experiment and expertise by enterprising Victorian gardening enthusiasts and have involved complex hybridisation of several varieties of smaller plums, including the native English bullace and sloe, but mainly larger and sweeter varieties from the Middle East and France. Among these were the damson and the cherry-plum which by the eighteenth century had become established fruits in Britain.

As the place-names indicate plums were often cultivated in gardens or orchards. As new and more succulent plums were introduced the wild fruits were virtually abandoned and are now usually found only in hedgerows and, occasionally, in the woods. Enthusiasts, however, still gather them to make sloe gin and bullace liqueur and even jam, although the damson and the modern plum are most peoples' preference.

Knowledgeable country-folk used to claim that bullace wine could hold its own both for taste and colour when compared with port: certainly Englishmen had acquired a taste for this particular drink well before the treaty with Portugal in 1703 which brought such alcoholic pleasure and tribulation to the upper classes.

Plums appear as items in Grasmere tithe payments in the Tudor period – that is before the modern varieties were likely to have been grown in Cumbria – so presumably the orchards of Plumgarth, Plumblands and Bullace Croft cultivated the bullace and the inhabitants of medieval times considered the sloe and the wild plum a welcome addition to their diet. Bullaces were certainly familiar to the household of Millom Castle in the early eighteenth century as they appear regularly in the domestic accounts of Madam Bridget Hudleston, the contemporary chatelaine.

In his 1997 *Flora of Cumbria*, Geoffrey Halliday stated that 'the bullace in Cumbria now rarely produced fruit.' If this is indeed so then this once well-known berry, larger than the sloe and rounder than the damson or plum, has been lost to the Cumbrian countryside.

# ROWAN TREE

*Matthiolus*

**Derivation:**   Old Norse: *raun*

**Place-names:**   Roanrigg, Roantrees, Roundthwaite (1256), Rauntreslak (1278), Rowantree Bank (1563), Rowan Park, Rowantree Beck, Rowantree Cove, Rowantree Crag, Rowantree Dub, Rowantree Force, Rowantree Gill, Rowantree Hill (1603), Rowantree Howe, Rowantree Knotts, Rowantree Sike, Rowantreethwaite, Roans

The rowan or mountain ash is well-known throughout northern Britain, its creamy white flowers filling the spring air with a rich perfume and its bright autumn berries shining brilliantly red against waterfall or crag. The rowan was believed to be a magic tree, more powerful than any other against witchcraft and elves and other manifestations of evil or mischief. Sprays of rowan were placed over the doors of houses, byres, sheepfolds and churches for protection against misfortune, disease or evil spirits; well-heads were dressed with rowan; milk churns had wreaths of rowan woven round them; and a sprig of rowan was used to stir the cream to make the butter come.

Those who were bewitched or barren could be restored to health and fruitfulness with the aid and protection of a rowan spray: it is from this belief that the tree received its other northern name, the Quicken (or Wiggen) tree, the tree of life. The rowan figures largely in the early rituals connected with the celebration of May Day, the great European festival of fertility and the coming of spring. In the 1760s Thomas Pennant was on tour in Cumberland and noted that 'till of late years the superstition of Beltain* was kept up in these parts and... it was customary for the performers to bring with them boughs of the mountain ash.'

The aura of magic surrounding the rowan has a long history. It features prominently in Norse mythology and is particularly associated with the goddess, Sif, wife of Thor and renowned for her long tresses of golden hair. The legend relates that Thor was rescued from a river by clinging to a branch of a rowan which Sif caused to bend over for him to reach. Henceforward the rowan was endowed with magic powers of protection against all manner of evil, witchcraft, and mischief from elves and other

* Beltane was the ancient name of the Celtic May-day rituals and spring festivities.

malign spirits. The red autumn berries were believed to be especially powerful in this respect: a cross of rowan twigs with red berries was placed over the doors of houses and byres and a simple rhyme declared that:

> *Rowan tree and red*
> *Leave the witches all in dread.*

The wood of the rowan tree was used for the staves or panels on which 'sacred' runes were carved and the famous Viking ships had one plank of rowan wood incorporated in their construction. Belief in the magic power of the rowan appears in so much of the folklore of the early medieval centuries.

Even so, the medieval country-man cast a practical eye on the rowan as he did on everything in the natural world around him. The actual wood was inferior to many others for most purposes but, as John Evelyn noted, it was very efficient for making bows. The berries may be very nutritious but there were others which were more pleasant to taste and less trouble to gather. On the other hand, they were a useful bait to trap birds easily and they were helpful in the treatment of a sore throat and to treat skin problems such as scurvy. Perhaps the main attraction of the rowan berry may have been in the inebriating spirit which could be distilled from its juices.

The rowan tree has a central place in the folk culture of all the peoples of the North and, although we no longer observe the ancient rituals or accept the ancient beliefs, it still holds its place as a favoured and most aesthetically pleasing tree. There must be few who do not pause in their walks in the fells to admire the rowan's graceful outline against waterfall or crag or clinging to the steep sides of a high ravine with its delicate white flowers in the spring or its brilliant red berries.in the autumn.

# SPINDLE TREE

*Matthiolus*

**Derivation:**  Old English: *spinele*
**Place-names:**  Spindle Bank, Spindle Wood, Spindle Head
(1614)

In the autumn the spindle is a pleasantly decorative tree with
its display of dark red leaves and orange-coloured fruits. For

the rest of the year it is rather a dull, inconspicuous shrub almost deserving the rather quirky names it has acquired in various parts of the country: skewerwood, prickwood, pegwood, louseberry and gattertree. The last was its usual name in medieval times and it was believed by many, including John Gerard, to mean 'the goat tree', derived from the Old English *gat* or the Old Norse *geit* a view apparently confirmed for Gerard by the fact that, as he alleged, 'this shrub is hurtfull to all things, especially to goates.'

It seems more probable that this older name originated in the Anglo-Saxon words *gad-treow*, the goad tree, as it was found that the tough, hard wood was especially suitable for making ox-goads in the days when oxen were in general use for ploughing and other farm work later done by horses. The name 'spindle-tree' was introduced by the sixteenth century botanist, William Turner, who noted that the Dutch called it the *spilboome* and used its wood to make their spindles.

Spindles, skewers, pegs, and goads were made from the spindle tree long before Turner's day, and we learn from Chaucer's *Canterbury Tales* that the medicinal properties of the spindle-berries were also well-known in medieval times. Pertelote's prescription for Chanticleer's ill-humour – a 'laxatyve' of 'gaitrys beryis' – for a day or two, was clearly part of the lore of fourteenth century medicine.

Many modern herbalists would agree that Pertelote knew her medicine and Gerard was in no doubt at all that she was right when he wrote that 'If three or fower of these fruits be given to a man they purge both by vomit and stool.' Pertelote may well have been familiar too with the fact that there was no surer way of ridding hair (and, presumably, feathers) of lice than a dusting with the baked and powdered berries of the louseberry tree.

Some, at least, of these uses for the spindle tree would have appealed to earlier generations of Cumbria but it seems unlikely that there would have been any local contemporary demand for spindle wood to make bows for violas and keys for virginals.

# YEW TREE

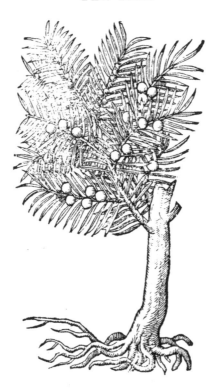

*Gerard*

**Derivation:**   Old English: *iw*

**Place-names:**   Yews, Yewbank* (several), Yew Tree Tarn, Yew Pike, Yewbarrow,* Yewthwaite, Yew Beck, Yew Crag,* Yewtree, Yewdale

---

* These names may possibly refer to ewes.

The yew tree is the oldest living thing on Earth. There are more than a dozen yew trees in Britain more than 4,000 years old and more than 50 which were here when the Romans came 2,000 years ago. Cumbria has an impressive number of these remarkable trees: the yews at Armathwaite Hall and St Martin's Church in Martindale are estimated to be over 2,000 years old and those at Lorton and Old Church are believed to have been growing there for about 1,500 years. Many more churchyard yews were here when the Normans arrived in 1066 – as, for example, those at Muncaster, Greystoke and Morland – and others dating from the early Middle Ages may be seen at Witherslack, Kirkoswald, Patterdale and Lanercost Priory. The yew is one of the few truly native trees of Britain and, together with the juniper and the Scots pine, is one of the trio of native conifers.

Long before Christianity arrived the yew tree was an object of veneration throughout Northern Europe. Celts, Angles, Saxons and the Nordic peoples planted yew trees by their burial mounds and other sacred sites and included them in their elaborate spring, autumn and mid-winter festivals. The yew tree's longevity, its evergreen foliage and its ability to renew itself held out the promise of eternal life. From this may have arisen the custom of placing small branches of yew in the coffins of the dead: in *Twelfth Night* Feste's song contains the lines:

> *My shroud of white, stuck with yew,*
> *O, prepare it.*

The red berries of the yew tree and their toxic seed offered protection against the evil spirits which were always waiting to do harm to humanity. There was also a widely held belief that if one peered through a cleft in a yew tree it was possible to see the spirit of the dead.

At the heart of the mythology of these people was the World Tree, Yggdrasil, the Tree of Life, usually assumed to be an ash tree, but it is referred to in the Norse language as *barraskr* which translates as needle ash with evergreen needles. The ash tree does not have needles, evergreen or otherwise, nor is the ash noted for its longevity. It seems probable that Yggdrasil was not an ash tree but a yew tree.**

Superstition, veneration and religious belief, however, went hand in hand with practicality in Man's relations with the Natural world and the yew tree had a number of practical uses. Its wood is hard, elastic and durable and was much used before iron came into general use. In particular it was the best wood

---

** Note on Yggdrasil

In Norse mythology Yggdrasil was the tree of Life, evergreen and eternal, whose branches covered the whole world. It had three roots: Asgard, the home of the gods; Jotunheim, the home of the giants; and Niflheim, where the dragon Nidhug lived constantly gnawing away at the roots of the tree which, being eternal, constantly renewed itself.

By a spring – Urd's well – in Asgard lived three sisters – the Norns – who watered the tree daily and who sat spinning the fate of all humanity. At the top of the tree sat an (un-named) eagle which carried messages throughout the world. The eagle and the dragon were bitter enemies.

Also at the top of the tree was a squirrel, known as Ratatosk, who carried messages between the dragon, Nidhug, and the eagle. These messages were usually provocative and insulting and designed to stir up trouble, so much so that eventually this is thought to have been linked to the full-scale conflict known as Ragnarok, the Götterdämmerung, the Twilight of the gods, the destruction of the Old World.

for yokes for draught oxen or for carrying pails of milk or water; it was also considered to be the most reliable wood for hunting spears. But by far its most common use was to make the longbows which featured so famously in English victories in the battles of medieval warfare.

In Shakespeare's *Richard II* Scrope informs the dejected King that:

> *The very beadsmen learn to bend their bows*
> *Of double-fatal yew against thy state.*

The yew tree forests of England and Wales were virtually destroyed to equip the archers who fought at Crecy, Poitiers, Agincourt and other conflicts of the Hundred Years' War and even into the Tudor period, for 139 yew longbows were recovered from Henry VIII's flagship, the *Mary Rose*, when it was raised from the seabed in 1982. Remnants of these yew forests may still be seen in many parts of the country, notably in the chalklands of the south but also in Cumbria's limestone areas at Arnside Knott, Whitbarrow and Haverthwaite Heights. Norman Nicholson wrote of a wood of them near Beetham, dark and Gothic, like a crypt with red pillars and blue-green fan-vaulting.

Cumbria's medieval inhabitants may have regarded the yew as a guardian of their household against evil spirits, bogles and various mischief-making goblins but they were also concerned to keep their children and livestock away from the seductive berries with their highly poisonous seeds. Whatever its other virtues the yew seemed to have no medicinal value for almost all parts – bark, needles, and berries are toxic and usually fatal if eaten by humans. 'Slips of yew' were among the many noxious ingredients in the witches' unappealing cauldron in *Macbeth.*

What could not then be known was that the yew tree held secrets only recently discovered by medical science. The bark and needles of the yew, apparently of little use to humanity, have been found to be a source of baccatin a direct precursor of the drug taxol which has had proven success in the treatment of certain types of cancer.

Fame came to a few Cumbrian yew trees not through any place-name nor through their practical use or superstitious belief but by historical accident, poetic inspiration and personal eccentricity. The Lorton Yew – Wordsworth's *Pride of Lorton Vale* – was famously used by George Fox as a pulpit when in 1652 he came to bring his Quaker message to a large gathering of dales-folk. The Borrowdale Yews inspired Wordsworth to compose one of his well-known verses – *The Fraternal Four* – and the Ordnance Survey awarded them the honour of being the only trees in Lakeland to be named on their maps. The impressive array of fifteen yew trees at High Yewdale were planted by the farmer there, a contemporary of Wordsworth's, to commemorate the birth of each of his children. And finally, the internationally famous yew tree topiary at Levens Hall we owe to the fashion for this type of artistic gardening which swept through Europe following the example of Louis XIV's palace gardens at Versailles.

# 4
## ANIMALS

## AUROCHS AND THE BISON

*Above, C. Hamilton Smith, copy of the sixteenth century 'Augsburg Aurochs' painting and, below, Thomas Bewick, a Chillingham bull*

**Derivation:** Latin: *ursus*
Old English/Old Norse: *ur*
Old German: *urochs*
**Place-names:** Urswick

Urswick is recorded as *ursewica* in the twelfth century, a name often interpreted as 'the dairy farm by bison lake' (*ur* + *sae* + *wic*) but this analysis may be misleading as the Old English/Old Norse word *ur* is usually taken to refer to the aurochs, a similar but genetically different animal from the bison. It is also relevant to note that the medieval word for the bison was the *wisent* (Old English *wisend*/Old Norse *visundr*) and that the word bison was not introduced into the English language until the seventeenth century.

Aurochs have been extinct since the last one died in a Polish forest in 1627 but it is known that they once roamed most parts of Europe except Ireland and the far north of Scandinavia. The most complete skeletons were found in Denmark in the twentieth century and are now in the Danish National Museum and form the basis for reconstructions of the animal which was a formidable beast described by Julius Caesar as fast-moving and aggressive. A bull aurochs stood up to two metres tall and could weigh up to a ton (c1000kgs). It was equipped with powerful horns often a metre in length, prized trophies for the hunter which he could bind with gold to form prestigious drinking horns. Several such horns were discovered in a chieftain's grave in Germany, including one elaborately gold-bound example which could hold five litres of wine. The Anglo-Saxon ship-burial at Sutton Hoo also contained a number of much less finely decorated aurochs drinking horns.

Bones of the aurochs have been unearthed in many parts of Britain as far apart as Caithness in Scotland and Porlock Bay in Somerset and also in Caermarthen in Wales and Urpeth in County Durham. In Cumbria remains of the aurochs have been found in several locations – at Burgh Marsh by the Solway Firth, near Silloth on the marshlands of Moricambe Bay, in the estuarine lands at the mouth of the River Irt near Ravenglass, and in the wetlands surrounding Sunbiggin Tarn – all typical grazing grounds of the aurochs, whereas the European bison (also known as the wood bison) prefers a forest environment.

No aurochs remains have so far been discovered at or near Urswick and this has led some linguists to put forward an alternative interpretation for this place-name. It is suggested that the derivation could be not from Old English/Old Norse *ur* but from Old English *ora* (*ars* in its genitive form), referring to the iron ore deposits nearby which, it is believed, were mined by the Romans. This would result in a meaning of iron ore settlement.

There can be no doubt, however, that, in the distant past, small herds of aurochs did roam the British countryside but they probably became extinct here long before their relatives in Europe. They were the direct ancestors of our domestic cattle and at the present time a Dutch enterprise, the Taurus Project, is engaged in genetic experiments which it is hoped will eventually lead to the successful re-creation of the aurochs which has been extinct in Europe for almost 400 years. It is now anticipated that the regenerated aurochs will be grazing and improving bio-diversity in the wilder regions of Europe by 2025.

# BADGER OR BROCK

*Bewick*

**Derivation:**    Old English: *brocc*
                   Dialect: *brock*
**Place-names:**   Brockhole, Brockbarrow, Brock Crag (several), Brockel Bank, Brock How, Brocklebank (1317), Brockle Beck, Brockley Bank, Brockshaw, Brock Stone, Broxty, Brocklewath (1441)

The name 'badger' is comparatively modern and began to replace 'brock' during the sixteenth century. It is believed that the modern name refers to the badger's distinctive white stripe – his 'badge' – but there seems to be no adequate explanation for the change of name. The numerous place-names in Cumbria which refer to the 'brock' suggest that it was once one of the more common animals to be found in the region.

For much of the Middle Ages it would seem to have gone largely unmolested by people although smoked badger ham did feature on the menu at royal and aristocratic banquets. However, in common with many other wild animals and birds, the badger was classified as 'vermin' by Statutes of Henry VIII and Elizabeth I and several centuries of persecution began. In Cumbria the parish records present a shameful catalogue of payments for 'brock heads' at one shilling a head: ten heads in Dacre in 1736, 73 in Kendal in the eight years between 1668 and 1676, most of them from Long Sleddale, a slaughter which was repeated year after year all over the country.

Badger baiting and bull baiting also became popular entertainments and innumerable badgers were done to death in this particularly barbarous manner and, despite the outspoken hostility of many enlightened figures of the time, attempts to have this and other forms of similar cruelty to animals prohibited by law came to nothing until 1835 when an Act of Parliament made badger baiting illegal. By this date the *Westmorland Gazette* had already informed its readers that the last badger caught in Westmorland was in 1823, and shortly afterwards an eminent local naturalist announced that 'Badgers are now extinct in the wild state in Cumberland'.

Fortunately, this pessimistic verdict was mistaken; the brock has survived in many Lakeland valleys and its numbers are again increasing. Unfortunately, there are still a few individuals who regard badger-baiting as an acceptable, if illegal, 'sport', condemned more than 200 years ago as one of the distinguishing vices of the lowest and basest of people, (although one commentator reported that even the greatest ladies enjoyed these spectacles).

Badgers today are not the favourite animal of the farming community who firmly believe that they spread tuberculosis among cattle and would perhaps welcome a revival of the

Tudor vermin Statutes. But to others the brock is a colourful, fascinating and harmless character, an object of affection, which enthusiasts still gather in the twilight to watch at Brock Stones, Brockhole and in many other places in the Lakeland woods where setts have been established, some for very many generations. For most of us the enduring image of the badger is that of 'Mr Badger' created by Kenneth Grahame in *The Wind in the Willows*, a kind, bluff gentlemanly figure who dispenses wise advice and so effectively comes to the rescue of Mr Toad when Toad Hall is invaded by the dastardly weasels.

# BEAVER

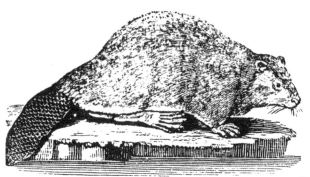

*Bewick*

**Derivation:**   Old Norse: *bjórr*
**Place-names:**  Barbon (1086)

The place-name 'Barbon' and an ancient beaver skull found in the same area form a modest indication that beavers may once have inhabited the rivers of Cumbria. Naturalists believe that, although a few may have survived into the sixteenth century, beavers became virtually extinct in Britain towards the end of the twelfth century and it even seems unlikely that there were many left at the time of the Norse settlements a little more than 200 years earlier.

This appears to be confirmed by two historical references from Wales. In about the year 940 Howel Dha tells us that a beaver's fur was priced at 120 pence while the skin of a marten was worth only 24 pence and that of a wolf, fox or otter no more than 8 pence, sure indications of rarity and abundance;

154

and in 1188 Giraldus Cambrensis states that beavers were then to be found only in the River Teifi.

Widely scattered place-names from other parts of the country suggest that at the time of the Saxon settlements the beaver was still fairly common: Beverley in Humberside, Bevercotes in Gloucestershire, Beversbrook in Wiltshire and Beverburn in Hereford and Worcester, all have their origin in the Old English word *beofor*, while Barbourne in Hereford and Worcester, Bardale in Yorkshire and Barbon in Cumbria are derived from the Old Norse *bjórr*.

The Domesday reference to Barbon as Berebrunna has led some to believe that the name may be derived from Old English *bera*, a bear, but there is no evidence that the brown bear which was present in Roman Britain survived very many years after the legions departed, much too early for the Norsemen and barely in time for the Angles and Saxons. The second element of the name Barbon – a burn or stream – also strongly suggests that it was the beaver and not the bear which frequented the neighbourhood when the Norsemen arrived.

In the eighth century the Christian Church placed a ban on the eating of beavers (in accordance with the dietary rules laid down in Leviticus II) and the few beavers that remained in the country were valued mainly for their fur which became an aristocratic luxury. There are certain references which suggest that in medieval times it was known that at the base of a beaver's tail is a scent gland which secretes an oily castor substance rich in salicylic acid, the basic ingredient of aspirin, for centuries the standard remedy for headaches. It is unlikely that this could have been very much used in Britain.

Beavers are no longer extinct in the wild in Britain. A small colony was discovered in the River Otter in Devon in 2014 and is now fully protected by law. They have also been re-introduced in several other counties in England and on a much larger scale in Scotland. Not everyone has welcomed this as beavers do have an impact on the landscape by building their dams on streams and by felling trees, but the naturalists maintain that the dams create wetlands which provide habitats for other wildlife and plants and that beavers do not fell large trees but encourage coppicing and so stimulate further growth. There are now several hundred beavers in Britain and it seems very probable that they are here to stay.

# Wild Boar

*Bewick*

**Derivation:**   Old Norse: *goltr, swin\*, griss\**
Middle English: *wilde bor*
**Place-names:**   Wild Boar Fell, Goatscar, Goatbusk (1574 -
lost), Swindale, Swine Crag, Swinegill, Swinklebank, Swin-
side, Swinstythwaite (lost), Swinstonewath, Grisdale, Grisee-
bank, Grisebeck (1331), Grisedale (1291), Grizedale,
Mungrisdale (1285), Swinsty

The wild boar which the Norse settlers found rooting in the
forests of Cumbria would have been a familiar sight to them
for the boar's head was a festive Yuletide dish throughout
northern Europe. Freyr, the popular god of peace and plenty

---

\*   These may also refer to the domestic pig.

of Scandinavian mythology, travelled the pagan world on a wild boar called Gullenbursti or Golden Bristles, the image of which fashioned into a helmet crest – 'The Swine of Battle' – was a Viking warrior's treasured possession.

The large number of place-names which apparently refer to the wild boar would suggest that this ferocious animal could be met in almost any forest glade. In Cumbria the vast Forest of Inglewood was described in the early Middle Ages as 'a goodly great forest full of woods, red deer and fallow, wild boar and all manner of wild beasts.'

Wild boar remains have been found in several Neolithic and Bronze Age caves indicating that it has an ancient lineage in Britain and it is also the wild animal with the longest historical record too. It appears on ancient British coins, in Celtic art, in many Roman inscriptions, in Saxon manuscripts and in numerous documents concerning feudal tenure and medieval Forest Law. As one of the three 'privileged' Beasts of the Forest (the wolf and the red deer were the others), the wild boar was protected by the Forest Laws, to be hunted only by the King or the socially exclusive few to whom he gave permission. Its image appears on the heraldic arms of several families of the nobility, notably that of King Richard III.

Hunting the wild boar could be a hazardous enterprise as many a feudal noble found to his cost when his 'Bore-spere' failed to stem the beast's violent onslaught. This was a danger well-known by Shakespeare's time for in his 'Venus and Adonis' when the handsome but naïve youth Adonis decides impulsively to go on a boar-hunt the justifiably alarmed Venus warns him of the danger:

> *O, be advised; thou know'st not what it is*
> *With javelin's point a churlish swine to gore,*

> *Whose tushes, never sheathed, he whetteth still,*
> *Like to a mortal butcher, bent to kill.*

And, sadly, the fair Adonis went off to meet his fate.

The life of the boar-hound, too, must have been exceedingly short if we are to believe a sixteenth century account of a boar hunt in which only twelve out of fifty hounds survived the day.

As the population increased and more and more land was put to the plough the boar's fate was doomed. The boar paid no regard to boundaries or cultivated fields and his onslaughts were devastating. Shakespeare's contemporaries would have fully appreciated his reference to

> *The wretched, bloody and usurping boar*
> *That spoiled your summer fields and fruitful vines*

Within a hundred years the wild boar was extinct in England, the last one, it is said, meeting its end at the hand of Richard Gilpin of Kentmere in the late seventeenth century when he finally dispatched a boar which 'had much indammaged the country people there'. A similar claim is made for Sir Richard Musgrave of Kirkby Stephen who is also credited with killing the last boar in England on Wild Boar Fell at about the same time, a feat 'proved' by the fact that a wild boar's tusk was found lying on his body when his grave was opened in 1847.

The Norse families who settled in Britain in the tenth and eleventh centuries would most probably have brought with them the custom of the Boar's Head as the centrepiece of their feast at the Winter Festival. This was in honour of their favourite god, Freyr, who rode the world on his boar, Gullenbursti. It became established as a Christian custom in royal and noble households in the Middle Ages and it is still a feature of

159

the Christmas festivities in several Colleges at Oxford and Cambridge where the boar's head, complete with an apple in its mouth and bedecked with rosemary, was ceremonially brought to the table accompanied by a trumpet fanfare. A special 'Boar's Head Carol' was sung for the occasion. Freyr and Gullenbursti disappeared and were replaced by St Stephen who is commemorated on 26 December.

All this was far removed from the concerns of ordinary farming families for whom the wild boar was an ever present threat, a dangerous and destructive enemy. The wild boar itself has recently been re-introduced into Britain, not yet in Cumbria but with some success in the New Forest.

# WILD CAT

*Bewick*

**Derivation:**   Old English: *catte*
                  Old Norse: *kattr*
**Place-names:**   Cat Bank, Cat Bells, Cat Bield, Catber, Cat Cove, Cat Crag, Cat Gill, Cat Hole, Cat How, Cat Keld, Cat Stone, Catsayle (1260), Catstycam, Cat Tails, Halecat

The wild cat is believed to have inhabited Britain since soon after the last Ice Age but now survives only in a few remote places in the Highlands of Scotland. A few centuries ago it was to be found in almost every part of Britain. The authors of the early Guide Books to the Lake District leave us in no doubt that in the second half of the eighteenth century the wild cat was fairly common in the woods and on the fell-sides.

Thus James Clarke in his *Survey of the Lakes* (1786) tells us that twelve wild cats were killed near Ullswater during

161

Whitsun week in 1759, adding that: 'The Wild Cats here are of different sizes, but all one colour (grey with black strokes across the back); the largest are near the size of a fox, and are the most fierce and daring animals we have; they seem to be of the tyger kind, and seize their prey after the same manner; they cannot be tamed, their habitation is among the rocks or hollow trees.'

A few years earlier in 1772 William Gilpin had found that the thickets among the Lakeland mountains were well frequented by the wild cat, and Thomas Pennant, travelling in the Windermere area at about the same time, commented that 'Wild cats inhabit in too great plenty these woods and rocks.' William Hutchinson's *History of Cumberland* (1794) relates that they were just as common on the fells of Eskdale and Wasdale.

Confirmation of these travellers' impressions is to be found in the Parish Records of Kendal, Orton and Barton from the seventeenth and eighteenth centuries. The payment for a 'Wilde catt head' was four pence in Kendal but as much as one shilling in Orton and Barton. Thirty-six wild cats were killed in Martindale between 1706 and 1739. With the development of enclosed sheep farming and the practice of breeding and preserving game, and especially grouse and pheasants for sport, it was clear that the depredations of this most fierce of British mammals could no longer be tolerated.

It is believed that the last one in Cumbria was killed on Great Mell Fell in the middle years of the nineteenth century. There were many who did not mourn its passing for it was a truly fearsome creature, willing and able to attack and inflict serious injuries on any human who ventured too close to its haunts. To the medieval farmer and his young stock in the

162

wooded valleys it must have presented a menacing hazard. The wild cat's prey included young deer and their fawns and most small mammals such as rats, mice, voles, and rabbits, but, given the opportunity, it was equally partial to poultry, lambs, calves and goat kids.

Shakespeare was clearly familiar with the wild cat. He referred to it on several occasions. The witches in *Macbeth* refer to its distinguishing stripe and its warning sound – 'Thrice the brinded cat has mew'd' – and in the *Merchant of Venice* Shylock, referring to the cat's habit of hunting by night and sleeping during the day, contemptuously comments to Jessica – 'He sleeps by day More than the wild cat'. It was well-known that the wild cat could not be tamed and a sustained punning reference is made to this in *The Taming of the Shrew* when Petruchio informs Kate:

> *I am born to tame you, Kate;*
> *And bring you from a wild cat to a Kate*
> *Comformable as other household Kates*

There would be few in the audience of the Globe who would not readily appreciate these lines for at that time the wild cat would be familiar to almost everyone. Place-names referring to it are to be found not only in the Lake District but in almost every part of the country from London to Lancashire and from Norfolk to Northumberland, and throughout Scotland and Wales.

# RED DEER

*Illustration – Dr Peter Delap*

**Derivation:**     Old English: *deor, heorot, bucc*
                    Old Norse: *dyr, hjortr, bukkr*

**Place-names:** Deer Bields, Deer Close, Deergarth, Deer-
ham, Deerscale, Deerslack, Hart Crag, Harter Fell, Hart Head,
Hart Rigg, Hartside, Hartsop, Buckbarrow, Buck Crag, Buck
Holme, Buck Pike, Buckscess, Buckstone

The red deer is the most magnificent of British native wild animals and by good fortune it is still with us in goodly numbers,
mostly in the Scottish Highlands but also in the New Forest,
on Exmoor, in the Brecklands and in the Lake District. The
splendid arrogance of the stag has made it a natural symbol of
power, strength and majesty: it appears in pre-historic Scandinavian rock-carvings, it decorates the ceremonial helmets of
Bronze Age chieftains, it adorns the royal standard from the
ship-burial at Sutton Hoo and it may be found in innumerable
coats-of-arms.

To hunt the red deer was the noblest of sports: William the
Conqueror declared that he loved the tall deer as if he were
their father and eagerly built on a custom begun by his Saxon
and Danish predecessors of creating 'Royal Forests', vast do-
mains within which the 'Beasts of the Forest' and the 'Beasts
of the Chase' became privileged species, reserved for socially
exclusive and bloodily ritualistic hunting. A seventeenth century Marchioness relates that it was customary for women and
ladies of quality after the hunting of a deer to stand by until
they are ripped up that they might wash their hands in the
blood, supposing it would make them white.

James Clarke, however, tells the story of a more practical
housewife in Martindale who acted promptly to take advantage
of the custom that the first person to seize hold of the hunted
deer could claim its head: she, for the sake of his head laid
hold of him as he stood at bay upon a dunghill, threw him

down, and getting upon his neck, held him fast. A free meal for her family was more important than white hands. This was the only legitimate way ordinary folk could obtain a meal of the 'king's venison', for the Forest Laws demanded severe punishment or even death just to be caught carrying a bow and arrow within the Royal Forests.

All this was in the centuries after the Norman Kings made the hunting of deer and feasting on venison a privilege exclusively for the nobility, and established in strictly enforced law what had been recognised in custom in earlier centuries. Organised hunting was not part of the way of life of either the Anglo-Saxons or the Norse people and, although deer bones have been found in their excavated villages both in Britain and in Scandinavia, they were present in only very small quantities compared with the remains of cattle, goats and pigs. These small family farmsteads were too busy clearing the forest for pasture and cultivable land and looking after their crops and livestock to spend time in hunting the large and elusive deer: they sought smaller, more easily captured game which were also considerably less trouble to prepare for the pot.

It was not the huntsmen but more than 400 years of deforestation and intensive sheep farming which ousted the deer from much of their natural habitat, and in Cumbria they survived only because of the 'protection' given to them in the parklands of local landowners such as the Sandys of Graythwaite, the Hasells of Dalemain and the Stricklands of Sizergh. Today the deer population in the Lake District has increased so rapidly that their numbers are becoming a matter of concern to foresters and farmers.

In Cumbria there are many place-names to confirm what archaeological and agricultural finds have plentifully revealed, that the red deer once roamed through all the Cumbrian forests, from Inglewood to Furness, from Copeland to Whinfell. The large number of place-names which honour the deer, the buck, the hart and the stag reflect a very different landscape and a very different society: the great forests have gone, and with them the harsh Forest Laws and the aristocratic tyranny they embodied.

# ROE DEER

*Bewick*

**Derivation:**    Old English: *raege*
                   Old Norse: *rá*
**Place-names:**   Raeburn, Rasett Hill (1224), Raygarth (1227),
Rayrigg, Rayseat (1224), Rayside (thirteenth century),
Rheabower Wood, Roegarth

The Roe Deer is Britain's second largest native mammal and
in the days when much of the country was covered with extensive forest it was among the most numerous of the larger
animals. Today it is found mainly in Hampshire and Dorset
and in northern England and Scotland. The widespread

destruction of the forests which occurred from the sixteenth century onwards deprived the roe deer of its natural habitat and it had almost disappeared by the end of the eighteenth century.

As a mere 'Beast of the Chase', and therefore inferior to the Red Deer which was a 'Beast of the Forest', the roe escaped the intensive hunting which brought other animals to extinction or near extinction but the economic needs of Man which accompanied the Industrial and Agricultural Revolutions achieved the same end. By the end of the eighteenth century the roe deer was extinct in most of Britain but in Cumbria the roe deer managed to survive, chiefly in the Solway Mosses from where in the previous century Lord William Howard was able to supply the table of King Charles I.

After a short period of rapid revival in the nineteenth century the roe again came under pressure from the loss of its natural browsing to grazing sheep and to the protective fences erected by the Forestry Commission to guard its new plantations. The efforts of modern conservation are now offering new habitats notably in Roudsea Wood, Claife Woods, Graythwaite and Grizedale Forest. It is difficult to arrive at an accurate estimate of the roe population for they rarely break cover and are usually very shy creatures. The Forestry Commission estimates that there are now about half a million roe deer in Britain and there are only a few areas where it is not found. Primarily a woodland creature the roe deer may now be found in open heathland and moorland.

Poaching the roe was a time-honoured country pastime, fraught with much less danger than pursuing the aristocratic red deer. The place-names tell us something of where hungry or adventure-seeking men of medieval Cumbria sought to sur-

prise the roe deer in the quiet crepuscular hours. One countryman in the early seventeenth century pleaded:

> *Is it not lawful we should chase the deer*
> *That breaking our enclosures every morn*
> *Are found at feed upon our crops of corn?*

Certainly, an early seventeenth century writer commented on
the partiality of the English for venison pasty, 'a dainty rarely
found in any other kingdom'.

Today the image of the roe deer for most people is not primarily as a wild animal of the woodlands or as the filling of a
pasty but as an adorable Bambi.

# FOX

*Bewick*

**Derivation:**    Old English: *fox, tod*
                    Old Norse: *refr*
**Place-names:**  Fox Bield, Fox Crag, Foxerth, Fox Fold, Fox
Gill, Fox Hole, Fox How, Fox Tarn, Foxyard,Todda, Todd
Crag, Toddell, Todd Fell, Todd Hills, Toddle, Tod Gill, Tod
Hole, Tod Rigg, Reagill

The fox did not become a serious enemy to man until the establishment of the great medieval sheep farms, the monastic
'Herdwicks'. Then his depredations among the young lambs
threatened the livelihood of all those who were involved in the
lucrative and prosperous wool trade. From the thirteenth century onwards the fox was beginning to replace the wolf as the
foremost enemy of shepherds, wool merchants, cloth mer-

chants, spinners, weavers and many   others whose fortunes depended on the humble sheep.

Earths and dens and other haunts were sought out and it is these which figure significantly in the place-names. In the new Charter of the Forest of 1217 the fox was named as a 'Beast of the Chase' and, as the centuries passed and the wolf and the wild boar became extinct and the deer were no longer readily found in every forest glade, fox-hunting steadily gained in popularity among the upper classes until by the late seventeenth century it had become the main sporting pastime of every country gentleman. In Cumberland fox hounds had been kept for King John as early as 1202 and the monks of Lanercost Abbey were given the right to one-tenth of all fox skins, a commodity rapidly increasing in value as the beaver disappeared.

As a creature of remarkable cunning the fox was often more than a match for his pursuers and soon earned a place in the literature and folklore of almost every country. In English literature the fox, either as Reynard or as Mr Tod, has featured as a principal character in innumerable stories, as, for example, in Geoffrey Chaucer's fourteenth century *Canterbury Tales*, in Ben Jonson's 1607 play *Volpone* or *The Foxe*, in John Harris's nineteenth century American tales of *Brer Fox* and *Brer Rabbit*, in Beatrix Potter's *Tale of Mr Tod* and in Richard Adams's *Plague Dogs*.

The fox is portrayed as cunning and intelligent, as deceitfully scheming and also as helpful to those planning to outwit mankind. But it was as a beast of wanton destruction in the farmyard and among the new-born lambs that in 1566 Parliament decreed that every parish had a statutory obligation to exterminate it, thus strangely contradicting the efforts being made by some landowners actively to preserve foxes on their estates to ensure a good hunt.

The statutory reward for a fox's head was one shilling but in the seventeenth century Sir Daniel Fleming of Rydal Hall was prepared to pay only sixpence. Better value was on offer at Hawkshead a few years later when the parish paid five shillings, equal to more than a week's wages for a farm worker then and for many years to come. Knowledge of the places where the fox's dens were to be found could be very rewarding.

Fox-hunting in Cumbria was rather different from the socially exclusive ritual of the southern counties; it was as much a necessity as a sport, and everyone took part. From the late eighteenth century organised hunts with packs of foxhounds became a feature of local life. Huntsmen such as Tommy Dobson of Eskdale and Joe Bowman of Ullswater became legends in their lifetime. The fox pursued in older days was not the familiar Reynard, the small red fox, but the grey fell fox, a formidable foe weighing up to fifteen pounds and capable of a six-hour chase of more than fifty miles. This was the fox which brought fame to John Peel, the 'tod' of all early place-names. It is now extinct.

Little is known of the fox in Anglo-Saxon and Norse times but one may assume that poultry had to be protected from marauders then just as they were in every other age. This is surely confirmed by the fact that these people were acquainted with the board game known as Fox and Geese which originated in northern Europe and was known as *Tæfl*. Thirteen geese have to escape capture by the fox but the fox too has to avoid 'capture' by the geese. An interesting glimpse into family life in the dark evenings on lonely farmsteads more than a thous
and years ago.

# HARE

*Bewick*

**Derivation:** Old English: *hara*
**Place-names:** Hare Beck, Hare Crag, Hare Gill, Haresceugh, Hare Shaw, Hare Wood, Harras

The hare is now scarce in Cumbria but it was once fairly common. From the thirteenth century we have a reference to the maintenance of ten harriers for hunting the hare and the early seventeenth century kitchen accounts of Lord William Howard indicate that a plentiful supply was forthcoming for his table. The place-name evidence is not always easy to interpret as it is often unclear whether the linguistic origin is from Old English *hara*, a hare, or from the Old Norse, *harr* meaning hoary or covered with lichen.

At least we know that we are concerned only with the native brown hare, a creature of the open fields and lowlands, and not with the white mountain hare since this was introduced only in the late nineteenth century and then only to Scotland and the Peak District

The hare is good to eat and it would be surprising if it did not appear on the tables of humbler folk as well as those of Lord Howard. And yet so persistent and widely believed were the legends and superstitions associated with the hare that it was outlawed by the early Christian Church – to eat its flesh was declared a sin in accordance with the dietary rules of Leviticus. Caesar tells us even before this that the Britons would not touch the meat of the hare as in most pagan myths it was regarded as a creature of the supernatural, ill-disposed to human beings.

Even its extraordinary leapings and cavortings in the springtime were considered to be some occult rites unknown to man. If the hares leaped in an anti-clockwise circle this was certain to foretell bad luck or a disaster. Medieval folklore embroidered all this with superstitious ingenuity: to meet a hare would surely bring misfortune or even death for a pregnant woman and it was likely to result in a deformed child, commonly one with a 'hare-lip'. It was also believed that hares were able to change their sex, to dance in time to music, to sleep with one eye open, to lay eggs at Easter, and to plunge those who eat its flesh into melancholy and madness.

None of this seems to have deterred those to whom hunting wild animals for sport was a laudable and almost ritualistic activity. The hare's astonishing speed and agility provided a very satisfying pursuit as Sir Daniel Fleming of Rydal Hall enthused after a day in the field in 1685.

There were, even then, voices raised in protest against the

cruelty of hare coursing, among them that of Edward Bury who, while accepting that it may be lawful to kill animals in order that we may eat, asserted that to sport ourselves in their death seems cruel and bloody. Most of the superstitions surrounding the hare had disappeared by the twentieth century but there can be no doubt that it was a noteworthy feature of medieval folklore creating anxiety and even fear especially among pregnant women.

Many of the superstitions associated with the hare have a long history. Druidical, Celtic, Anglo-Saxon and other pagan cultures all had rituals and beliefs linked to the hare's strange habits which were regarded as suspiciously different from other animals and were fraught with ill-will to humanity. Nor did the hare demonstrate any of the admired qualities of mankind; on the contrary it was regarded as timid, faint-hearted, totally lacking in courage. In Shakespeare's *King John* Philip the Bastard taunts the Duke of Austria with the words:

> *You are the hare of whom the proverb goes,*
> *Whose valour plucks dead lions by the beard.*

The hare achieved everlasting culinary fame with the implausible instruction which allegedly appeared in an eighteenth century cookbook by Hannah Glasse, entitled *The Art of Cookery made Plain and Easy*: 'First catch your hare.' The author actually wrote: 'Take your hare when it is cased' (i.e. skinned).

The more famous phrase first appeared in 1300 in Henry Bracton's *Laws and Customs of England* as: 'First catch your deer and, afterwards, when it is caught, skin it.'

# HEDGEHOG OR URCHIN

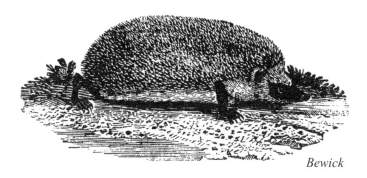

*Bewick*

**Derivation:**     Middle English: *urcheon*
**Place-names:**   Urchin Close, Urchin Rigg, Urchin Wood

The hedgehog is one of the oldest inhabitants of Britain. In Cumbria its fossil remains have been found alongside those of wolves and bears. It has few enemies – most predators are quickly deterred by its armoury of some 5,000 spines – but the hazards of hibernation and winter cold kept the population under a natural control. As Britain's only spiny mammal it attracted a good deal of interest and was regarded as an intelligent creature, the very emblem of craft and cunning.

In the Middle Ages, however, hedgehogs were roasted and eaten when other meat was scarce but in the eighteenth century they were kept as pets. The hedgehog's prickly coat made it so different from other animals that it became an object of suspicion in a world much given to superstitious beliefs. Perhaps the most persistent of these was the widely held belief that hedgehogs sucked the milk from cows. This was one reason why in the 1566 Act for the extermination of 'vermin' the hedgehog was included together with such ferocious predators as foxes, polecats and weasels among the animals whose extermination would be beneficial to mankind.

Cumbrian parish records show that this statutory obligation was carried out with conscientious thoroughness: one seventeenth century record indicates that the going rate for a dead hedgehog was two pence a head. In 1661 one parish 'disburst to Robert Whitehead for 17 hedgehogs £00.02s.10d.' This entry also reveals that by this date the old name of 'urchin' had already passed out of regular use although Shakespeare uses 'urchin' many times and 'hedgehog' only rarely. Roast hedgehog remained on the list of meats considered suitable for human consumption well into the eighteenth century.

Shakespeare's 'thorny hedgehog' was not to be seen and not to come near our Fairy Queen but Beatrix Potter's Mrs Tiggywinkle was clearly a more lovable creature who went to much trouble to wash and return Lucie's lost handkerchiefs. Perceptions of some of the inhabitants of the natural world have changed greatly in the past 400 years.

# THE MOLE

*Bewick*

**Derivation:** Middle English: *molle*
The dialect term *Moudy Warp* or *Mouldy Warp* is derived from the Old English *molde* or Old Norse *mold* meaning 'earth' or 'soil', and Old English *weorpan* or Old Norse *verpa* meaning 'to throw', with obvious reference to the mole's habit of creating molehills.

**Place-names:** Moldhylles, Moudy Mea

Mouldy Warp has long been a character in country folklore and as a lovable and unworldly participant in children's stories. The opening pages of Kenneth Grahame's *The Wind in the Willows* gave the mole a memorably unique place in English Literature, as, suddenly overwhelmed by the spirit of spring, he abandoned his brush and pail of whitewash and with a joyful cry of 'Up we go! Up we go!' he burst forth into the sunshine

179

and embarked on the remarkable adventures which culminated in the Battle of Toad Hall which brought immortal fame to 'the famous Mr Mole'.

Before Grahame's inspired tale the mole was generally looked upon as a rather unlovely creature which lived underground and created havoc by night on meadows and carefully tended lawns as he tunnelled his way in search of his unsavoury diet of slugs, earthworms and insect larvae. It was this which earned the mole the country name of mouldy warp or the 'earth thrower', a name as ancient as the Saxons and Norsemen who gave it to him. And it was this, and possibly the specific reference to the mole in Leviticus as 'unclean', which earned it a place in the 1566 Statute for the extermination of 'vermin'.

Landowners in medieval times must have looked upon the mole and its molehills as a serious problem in their fields as a comment by Shakespeare's impetuous Henry Percy (Hotspur) would seem to imply:

> *Sometimes he angers me*
> *With telling of the moldywarp and the ant.*

For four hundred years mole catchers in every parish did their utmost to achieve the purpose of the Elizabethan Act. In one parish in 1732 it was reported that 'no fewer than 5,480 moles had been killed'. In the mid-twentieth century more than a million moles were trapped every year in Britain. But this was not because they were still regarded as 'vermin' but mainly to satisfy the demands of a fashion for moleskin garments. The later development of synthetic fabrics put an end to this except for an occasional demand for moleskin trousers long known for their warmth and toughness.

The 1566 Statute authorised the payments to the mole catcher of three pence an acre except, that is, for any irreverent moles which burrowed in a churchyard. The reward for trapping these sinners was increased fourfold to one shilling, as recorded in Martindale in 1826. So thorough was the persecution of the mole that in 1797 the Agricultural Survey reported that we scarce ever saw a molehill upon the enclosed grounds of most of Cumberland. More than two centuries later many of the fields of Cumbria are again adorned with molehills just as they were when Moudy Mea – the meadow full of molehills – was given its name. Today farmers still trap moles and the mole also has to fear the silent swoop of a tawny owl.

# FIELD MOUSE

*Bewick*

**Derivation:**   Old English and Old Norse: *mus*
**Place-names:**   Mousegill (1370), Mousethwaite (1279), Mowsseaga, Muserg (1235), Musgrave (1202)

The field mouse – or wood mouse – is one of the oldest of British mammals. Ancient caves have yielded its Neolithic remains and for most of the time since then the woodlands and moorlands of Britain have provided favourable habitats. For several thousand years field mice lived out their short and harmless lives with only a few natural predators such as nocturnal owls to fear.

Occasionally, their colonies attracted man's attention but, for the most part, the field mouse did little to warrant human hostility. It is, therefore, surprising to find this tiny, inoffensive inhabitant of the woods listed alongside polecats, foxes and weasels in the Elizabethan Statute of 1566 which required churchwardens to organise the extermination of 'vermin'.

In the following centuries thousands of small animals were

182

done to death for reward and for sport. The national attitude to wildlife is revealed in Swift's *Gulliver's Travels* where we learn that Gulliver's greatest fear from the giants in Brobdingnag was that 'they would dash me to the ground as we usually do any hateful little animal that we have a mind to destroy.' This persecution and the destruction of the forests have greatly reduced the field mouse population – and, by remorseless ecological process, also its chief predator, the barn owl. Fortunately our woodlands are now being protected and replaced, and both the wood mouse and the owl now seem to have a more certain future.

Beatrix Potter assured the mouse of a young fan club with her tales of *Johnny Town Mouse*, *Mrs Tittlemouse* and the *Two Bad Mice*. Thus 'mouse' is almost restored as a term of endearment, even if a little barbed, as it was in Shakespeare's day when in *Love's Labour's Lost* Rosaline, in the course of a bantering exchange with Catherine, comments:

> *What's your dark meaning, mouse?*

It was not the wood mouse which so plagued the medieval housewife but the house mouse. These were the pests which had their tails cut off with a carving knife, so that, for a short while no doubt, as Puck said in *A Midsummer Night's Dream*

> *Not a mouse*
> *Shall disturb this hallowed house.*

Sprays of the tansy or branches of elder leavers were strewn about the larder and among the bed linen as a deterrent, a wise precaution, not merely on hygienic grounds but also because if a mouse ran over your foot it was a sure sign of misfortune to come.

# OTTER

*Bewick*

**Derivation:**     Old English: *oter*
**Place-names:**   Otter Bank, Otter Bield (1578), Otter Island,
Otter Keld (1581), Otter Stones, Otter Wood

The otter is a shy, sleek and graceful creature, a powerful
swimmer with a keen appetite for fish and a lively aptitude for
playfulness. Quiet, undisturbed waters by rivers, lakes or the
coast are essential to his way of life and in more peaceful and
less populous times the Cumbrian countryside was an ideal
habitat. Just over one hundred years ago the naturalist H. A.
Macpherson commented that 'It would be difficult to name
any part of Britain as better adapted to the taste of the otter
than the English Lake District.'

In his day otters were obviously flourishing throughout the area and, with a Victorian gentleman's innate enthusiasm for hunting and killing wildlife, he relates the exploits of the Kendal, Carlisle and West Cumberland Otter Hounds which killed hundreds of otters every year on the rivers of Cumbria from the Irthing to the Lune, from the Esk to the Eden.

Otter hunting has a long history and Acts of Parliament from the reign of Edward I to the later years of Elizabeth I had to be enacted in order to protect the otter during the breeding season, so serious were the inroads already being made on a never very large population. Whether the barbed spear of the medieval huntsman is preferable to the savaging by otter hounds of more modern times is arguable.

It is interesting to note, however, that it was not until the late eighteenth century that the persecution of the otter was cloaked in a form of moral justification. 'These nocturnal thieves', 'this midnight pillager' deserved only 'the sharpest vengeance' and to be brought 'to just disgrace'.

Such righteous ardour made catastrophic inroads on the remaining otters of England and Wales and even in the mid-twentieth century over one thousand otters were killed by otter hounds every year. Additional casualties were caused by industrial and agricultural pollution of the rivers. Otter hunting was finally made illegal in 1978 and this and belated conservation may save the day but it is salutary to reflect that at one time it seemed likely that all that would remain of the otter in Cumbria would be a handful of place-names.

It is a little surprising to find the otter included in a seventeenth century list of creatures considered suitable for human consumption for the 1566 Act had classified the otter as 'vermin' with a price on its head, a strong enough incentive for the penurious countryman to seek out the holts hidden in riverside banks, not to satisfy his hunger but for financial reward.

## PINE MARTEN OR SWEET MART

*Topsell*

**Derivation:**   Old English: *mearth*
                Old Norse: *murthr*
**Place-names:**  Mart Bield, Mart Close, Mart Crag (numerous), Marther Gill, Mart Knott

The Pine Marten, often known as the Sweet Mart to distinguish it from the evil-smelling Foul Mart or Polecat, has inhabited the Cumbrian woodlands since prehistoric times, once the second most common carnivorous wild animal in Britain. It is an attractive animal of extraordinary agility, able to climb a tall tree in seconds, chase a squirrel through the highest branches and fall sixty feet to the ground none the worse for the experience, and often with its prey in its clutches.

    Its chief predator is man and he has hunted the pine marten for sport and for its rich, dark brown fur for centuries. The medieval Forest Law included the marten with the roe deer, the fallow deer and the fox as one of the 'Beasts of the Chase' but its elusiveness and its habit of hiding in the treetops must have made it a poor day's sport. In Sir John Manwood's *Treatise and Discourse of the Lawes of the Forrest* it is unequivocally declared to be only 'the fowerth beast of the chase'. Certainly it escaped the general massacre of wildlife authorised by the 'vermin' Statute of 1566 and there are few references in parish records of payments for its head. This many explain why Thomas Machell in the course of his travel through the Lake District in 1692 could note that in Patterdale 'They have great store of Marts hereabouts and in Langdale Marts and wild cats are found all over.'

It was the invention of the sporting gun and the steel trap which put the pine marten in danger and for two hundred years in the eighteenth and nineteenth centuries, gamekeepers, country gentlemen and fur-hunters slaughtered it for sport, pleasure and economic gain. Martens have never been abundant in Britain in modern times and as they do not breed until their fourth year of life and then produce only two or perhaps three young each year this sustained onslaught quickly brought it near to the point of extinction. It is now rarely seen in England (although there were reports of a sighting in Shropshire in 2015) but there are still several thousand in Scotland and a few in North Wales. The shelter and protection of Forestry Commission enclosures and conservation reserves offer hope of a recovery and the pine marten's predatory hostility to the grey squirrel has also brought it firmly back into favour.

# POLECAT OR FOULMART

*Bewick*

**Derivation:**   Old English: *fulmart*
**Place-names:**   Foul Mart, Foulmart Dale, Foulmart Fold,
Foulmart Hill, Foulmart Railing, Foumart Gill

This fierce, aggressive and fleet-footed animal with its distinctive and unpleasant smell, known throughout Cumbria as the Foulmart or Foumart to distinguish it from the Pine Marten or Sweet Mart, was once found everywhere from the mosses of Morecambe Bay and Solway to the valley woodlands and farming country of central Lakeland.

Its usual diet consists of field mice, voles, frogs and any birds it can catch but it always regarded poultry and the inhabitants of game preserves as a bonus. Chaucer, in his *Pardoner's Tale,* refers to a polecat that was in his yard that had killed his chickens. It has been described as one of the most noxious

pests to which the farmyard is liable, the deadly enemy to rabbits, game and poultry. These incursions into man's territory
brought about its inclusion in the Statute of 1566 for the extermination of vermin but, in all truth, the polecat was persecuted more for its fur than as a threat to the food supply. The
Kendal Parish Records detail the course of its destruction in
the eighteenth century: '51 mart heads paid for at Easter 1702,
and 173 in 1794, and a mean average for most years in between.'

By the late sixteenth century when Shakespeare was writing *The Merry Wives of Windsor* the reputation of the polecat
was so widely regarded as morally reprehensiblc that it had
become a recognised form of insult as when Ford berates Falstaff (disguised as a woman): 'Out of my door, you witch, you
rag, you baggage, you polecat, you runyon! Out! Out!' The
foul smell of the polecat did nothing to diminish this reputation.

The predilection of the Georgian and Victorian country
gentleman for shooting any wildlife on sight in the name of
sport ensured an almost irreversible decline in the numbers of
foulmarts in all parts of the country. In Cumbria one enthusiast
boasted that he had personally shot 250 in the Lake District
over a period of 25 years. A fashion vogue for muffs and wraps
made of polecat fur made the steel trap a worthwhile invest-
ment and this finally condemned the polecat to extermination
in England. It lives on in Wales and attempts were made in the
1980s to reintroduce it into Grizedale Forest. Otherwise, at
present, all that remains of the polecat in Cumbria is half-a-
dozen place-names and accounts of exciting 'Mart-Hunts' to
be found in journals such as the *Westmorland Gazette* and *The
Field.*

# RABBIT

*Bewick*

**Derivation:**    Middle English: *coni*
**Place-names:**    Coneybeds, Coneygarth, Connyside, Cony-
pot, Cunnigarth, Cunning Garth, Conyhill (1371)

The word 'coney' to refer to the rabbit is now almost obsolete mainly because its original pronunciation rhymed with 'honey' or 'bunny' and this had associations which some later generations found unacceptable. Even in Shakespeare's day it was considered to be rather 'naughty' as the bard himself demonstrated in an exchange between Rosalind and Orlando in *As You Like It*.

By Victorian and Edwardian times *Bunny Rabbit*, *Brer Rabbit*, *Little Grey Rabbit*, and *Peter Rabbit* had all become established as part of a happy childhood and had created an image of the coney totally different from that known to earlier generations. Even the new dimension given to the rabbit world by the disturbing realism of Richard Adams's *Watership Down* bears little relation to the severely practical necessity of increasing the food supply which brought the rabbit into this country.

Precisely when this occurred has been a matter of minor controversy. The discovery of a rabbit skeleton in Norfolk in 2005 which was dated to the age of the Roman Occupation of Britain was considered by some to indicate that the rabbit was brought here by the Romans. However there is no reference to the rabbit for a thousand years after this and Domesday Book, which overlooked very little, does not mention it. The first references appear in the twelfth century from the Isle of Wight and the Scilly Isles, but by 1251 the rabbit was clearly a very special item on the royal menu as 450 were served at Henry III's Christmas feast in that year. Together with hares, pheasants and partridges the rabbit was classed as a lowly 'beast of the warren' and it was to be many years before it was legally on the tables of ordinary country folk.

So, it would seem that the Norman feudal aristocracy discovered that the English agricultural economy was incapable of providing an adequate supply of meat throughout the year and imported the rabbit to fill the gap. They had already discovered the rabbit in their Mediterranean adventures and it was not long before the new rulers of England had appointed 'warreners' on their new lands, men with status and authority, whose work it was to create warrens to ensure a supply of good quality meat for the lord's household and to see that the peasant did not poach or trespass.

A Statute of 1399 forbade 'labourers and artificers' to enter private warrens and as late as 1816 a further Statute condemned any 'labourer' caught poaching rabbits to seven years transportation. Fortunately it is notoriously difficult to keep rabbits within any defined area and long before the end of the Middle Ages the prolific coney had itself broken bounds and set up warrens on common lands everywhere. The place-names probably represent only the private warrens which by the nineteenth century held just a fraction of the total rabbit population.

# RAT

*Bewick*

**Derivation:**  Middle English: *ratoun*
Local dialect: *ratten*
**Place-names:**  Ratten Haw, Ratten Mire, Ratin Row (1366),
Ratten Row, Rattenrawe Lane (1279)

The Old English or Black Rat probably arrived in England
from Asia soon after the Norman Conquest as trade with the
Mediterranean lands steadily developed. The cooler climate
of Britain led it to seek a sheltered environment here and so it
became a creature of the medieval towns where it flourished
and eventually became an unpopular pest, while in the countryside it made its home in the warmth of byres and barns or
even in the woodwork of cottages, and here, too, its partiality
for grain and the food in the housewife's larder made it an unwelcome guest.

Grain stores had to be raised on tall stone 'toadstools' in an attempt to thwart the depredations of this very agile raider. Together with the frog – a foul and filthy creature – and spiders which were so loathsome that ladies screamed at the sight of them – the rat was of hateful aspect, a mischievous and destructive vermin. Nor was the black rat's reputation enhanced by the belief that it was responsible for the outbreaks of the bubonic plague which afflicted Europe at intervals throughout the Middle Ages.

The Black Death of the mid-fourteenth century is believed to have claimed the lives of more than one-third of the three million population of England. There is now some doubt as to whether the black rat flea was, in fact, the main carrier of the disease and, in any case, this was not a question which troubled the minds of those who faced the perils of the Great Plague: to them it was brought about by the wrath of God, a divine punishment for the sins of Man.

It is no surprise to note that the rat was included in the long list of 'vermin' to be destroyed by order of the Statute of 1566. It was left to a later age to point out that outbreaks of the plague ceased with the virtual extermination of the black rat by the larger, more aggressive and incredibly prolific brown rat which arrived in Britain in the eighteenth century. A few colonies still survive, usually on islands, but the black rat is almost certainly extinct in Cumbria and these few medieval place-names are all that remain to remind us that it once passed this way.

# RED SQUIRREL

*Bewick*

**Derivation:**    Old Norse: *ikorni*
**Place-names:**  Ickenthwaite (1539), Icornshaw (1279)

The word squirrel – from the Anglo-Norman *esquirel* – first appeared in the fourteenth century in Chaucer's *Romance of the Rose* and gradually replaced the Norse *ikorni.*

Two Norse place-names and two squirrels carved on the Bewcastle Cross are the only evidence we have that these animals have lived in the Cumbrian woods for a very long time. We do know that the red squirrel was one of the last mammals to arrive in Britain before it was cut off from the mainland of Europe about 8,000 years ago. We also know that the Scots pine and the hazel and the oak from which the squirrel obtains a substantial part of its diet were flourishing throughout the

196

new island soon after the ice had gone.

The great forests of medieval Britain provided ample food and shelter, and with few predators able to catch it, the squirrel enjoyed a long period of peaceful existence free from harassment and persecution. But man soon recognised this agile creature of the woodlands as an unusual, rather quirky, character of the natural world with its noisy chatter and secretive habits. Norse mythology found an appropriate role for the squirrel which, under the name Ratatoskr, was employed to convey messages and gossip from the eagle at the top of the World Tree, Yggdrasil, to the dragon who lived among its roots.

Shakespeare indicated in *Romeo and Juliet* that the squirrel also had a role as coachmaker to Queen Mab, the Queen of the fairies whose

> *chariot is an empty hazelnut*
> *Made by the joiner squirrel or an old grub,*
> *Time out o' mind the fairies' coachmakers*

The squirrel was neither a danger or useful to people, nor was its behaviour at all morally reprehensible, so it was largely ignored for most of the medieval period although it did appear in a list of animals considered suitable for human consumption. But the strangely inconsistent attitude towards creatures of the natural world which prevailed in the eighteenth century meant that squirrels (and many other wild creatures) were either kept as pets or were ritually slaughtered at Christmas and New Year. The naturalist Edward Topsell commented that squirrels were 'playful, sportful beasts... very pleasant playfellows in the house'.

The 'vermin' Statute of 1566 did not specifically name the squirrel but it did unleash an unprecedented destruction of wildlife whether named in the Act or not. As the human population increased and as agricultural and industrial changes took place the onslaught on the forests became ever more widespread. Nostalgic Elizabethans were already looking back on the days when 'this whole country's face was forestry'.

A century later Gregory King believed that no more than eight per cent of England and Wales was forest or woodland; by 1800 another million acres had gone, and by the beginning of the twentieth century Britain had the smallest amount of woodland in Europe. Steadily deprived of its habitat the red squirrel declined in numbers during these years; and there were other hazards too. The arrival of the grey squirrel in Britain in the 1870s has now presented the native red squirrel with its greatest challenge for survival for wherever the grey squirrel has established a foothold the red has disappeared.

The Forestry Commission estimates that there are now only 15,000 red squirrels surviving in England and fewer than 150,000 in the whole of the United Kingdom whereas there are believed to be about 2.5 million greys. It was not without good reason that Beatrix Potter (and others before her) depicted the red squirrel forever travelling in search of a new safe haven. Will the native squirrel vanish from the Cumbrian woods as completely as his Old Norse name, *ikorni*, now preserved only in the language of place-names? The introduction into the Lakeland scene of the grey squirrel's chief predator, the attractive pine marten, might save the day.

# WEASEL

*Bewick*

**Derivation:**   Old English: *wesule*
**Place-names:**   Weasdale (1581), Weasel Gill

To those of our own and earlier generations given to attributing human qualities to different species of animals the weasel has always been a creature of reprehensible character. The fox may be cunning, the goat lustful and the snake treacherous but the weasel is for ever cruel, voracious and cowardly, and, as Pisanio informs Imogen in Shakespeare's *Cymbeline*, notably quarrelsome.

For a tiny animal less than ten inches (25cm) long and weighing no more than four ounces (100gm), and still able to give a hard fight to a predatory cat or buzzard, this seems a harsh judgement but, even so, it accords with the Statute of 1566 which officially classified the weasel as 'vermin' to be exterminated in every parish in the land. Or, equally the explanation may lie in the weasel's tremendous appetite for mice – several hundreds in a year – and the mouse has been the darling of the children's nursery rhymes and stories for countless generations.

The farmer has an ambivalent attitude to the weasel: it is his ally in his battle against the grain-devouring mouse and vole, but an ally quickly becomes an enemy when it slaughters his newly hatched chickens. In *As You Like it* Shakespeare refers to the weasel's well-known reputation for raiding the hen-houses to suck the eggs, one of the most important items in the family's daily diet.

The persecution of the weasel is now far less widespread than in the hey-day of the game-keeper, and its numbers are once more increasing. Our two place-names indicate a fairly significant colony in those particular spots but they were latecomers, for the remains of their ancestors were found in the Neolithic layers of Dog Holes Cave on Warton Crag a few miles away.

# WOLF

*Bewick*

**Derivation:**   Old English: *wulf*
Old Norse: *ulfr, hvelpr*

**Place-names:**   Wolf Crags, Wolf Gill, Wolf Howe, Wolf Lea, Wofa Holes, Woffecowe, Woof Crag, Woof Gill, Woofergill, Wooloaks, Wuffetgarth, Hullockhowe, High Ouvah, Owshaw, Uldale (1230), Ulfshaw, Ulgraves, Ullathorne, Ullock (1230), Ullscarf, Ullstone, Ullsmoor, Ullthwaite, Ulpha (1337), Ulphanrigg, Ulsber, Ulvergill, Ulversdalebeck, Whelpeth (1452), Whelpo, Whelpside, Whelpstruther, Whelpsty Gill

It is not always possible to distinguish between place-names derived from ON *ulfr*, a wolf, and those derived from the Norse personal name *Ulfr*. Among the latter are Ullswater, Ulverston and possibly Uldale.

The demise of the bear in Britain, probably sometime in the sixth or seventh century, left the wolf as the most formidable predator for the medieval sheep-farmer and the woodland herds of deer to contend with. Although it was rarely as fierce a beast as popular legend often portrays it, the wolf was undoubtedly a serious and constant threat to the precious livestock of the early homesteads and, later, to the large sheep flocks of the monastic estates.

As the many place-names indicate the wolf was a very familiar beast to the English and Norse settlers and in 1016 the Forest laws issued by King Cnut declared that 'wolves were neither reckoned as beasts of the forest nor of venery therefore whoever kills any of them is out of danger of forfeiture'. This acknowledgement of the problem posed by the wolf population helps to explain why fifty years earlier King Edgar decided that the gold and silver penalty imposed on the defeated King of Wales should be changed to a yearly tribute of 300 wolf skins, an annual payment which lasted until after the Norman Conquest.

Written records from the years of the Norman kings leave no doubt that wolves were present in most parts of England in such numbers as to be a serious threat not only to the economy of the country but even to human life: a sanctuary was established in Yorkshire as a refuge for travellers who were threatened with an attack by packs of hungry wolves. So urgent did the problem become that it was not unusual for grants of land to be made with an obligation to control the local wolf population. Robert d'Umfraville in the late eleventh century was under such an obligation to protect Redesdale against the wolves. Even convicted criminals could be pardoned in return for producing a stated number of wolf tongues within a given time.

There are numerous places throughout northern England where wolf remains have been excavated and where the place-names clearly refer to former dens or haunts. In Cumbria wolf bones have been found near Kendal and Grange while the wolf appears in local place-names more frequently than any other animal. There are no fewer than six Uldales (valleys plagued by wolves), several Ullthwaites (forest clearings haunted by wolves), a number of wolf-hills (Ulpha, Ouvah, Wolf Howe), and some fifty other names referring to the wolf and its cubs or whelps.

We learn from a thirteenth century *Book of Husbandry* that in England at that time shepherds had to guard their flocks by night and that, during the daytime, they walked with a wolfhound in front of their sheep, with more wolfhounds following on behind, to protect them from marauding wolves lying in wait. Many years of sustained effort to control this ubiquitous hazard to sheep farming meant that by the fifteenth century its numbers were so depleted that the wolf was elevated to the status of a 'Beast of the Forest', the pursuit of which was reserved for the nobility only, with royal permission. Even so, a century later during the reign of Henry VII, the wolf was extinct in England.

As in other parts of the country Cumbria has several places where 'the last wolf was killed': the most common claimant is Humphrey Head where, according to the *Annals of Cartmel*, Sir John Harrington performed this feat to prove his love for his lady. An equally plausible piece of folklore has it that the caves of Glaramara were the wolf's last stronghold in England while others maintain that the Forest of Bowland or the Derbyshire Peak may also claim this dubious privilege. Whatever the truth may be, the many place-names alone clearly suggest that to the medieval inhabitants of Cumbria the wolf was an all too familiar beast whose haunts and lairs were well-known both by name and reputation, a wild beast which was feared more than any other, not least for its cunning, sinister, furtive nocturnal attacks on its prey – a fear expressed by Shakespeare in Tarquin's assault on Lucrece:

> *Alarmed by his sentinel, the wolf,*
> *Whose howl's his watch, thus with his stealthy pace,*
> *With Tarquin's ravishing strides, towards his design*
> *Moves like a ghost.*

The enduring fear and threat of the wolf felt by earlier generations may be seen in the number of phrases long embedded in the English language: e.g. a lone wolf, to cry wolf, to keep the wolf from the door, a she wolf, a wolf in sheep's clothing, a wolf pack, to throw someone to the wolves.

# 5
## DOMESTIC ANIMALS

The pastoral economy of the English and Norse settlers in Cumbria during the pre-Conquest and early medieval centuries seems to have been based largely on cattle, pigs and goats with only a smaller number of sheep. The chronology of the relevant place-names confirms what other sources also suggest – that the changes which led to the later predominance of sheep were directly due to the economic advantages brought by the immensely lucrative trade in wool and textiles in the fourteenth and fifteenth centuries and by the development of extensive sheep runs throughout the north by powerful monastic foundations such as Fountains and Furness.

More than two-thirds of all the place names relating to domestic animals refer in some way to either cattle, pigs or goats while a significant proportion of the 'sheep' names date from the late medieval period or after. The earlier 'cattle' and 'pig' names may have remained in use even when the land was turned over to sheep. Other livestock found in the place-names in significant numbers include the horse and the goose.

*Bewick*

# CATTLE

*Bewick*

**Cow**
**Derivation:** Old English: *cu*
Old Norse: *kyr*
**Place-names:** Coulandes (1295), Cow Close, Coa Garth, Cow Stead, Cowsty, Cowrake.

**CALF**
**Derivation:** Old English: *calf*
Old Norse: *kalfr*
**Place-names:** Calf Close, Calfgarth (1285), Calgarth, Calthwaite, Calva, Calvaload (1422), Caw Fell

**HEIFER**
**Derivation:** Old Norse: *kviga*
**Place-names:** Quypot (1279), Quiepotte (13th century), Wyegarth

*Bewick*

## Ox
**Derivation:**    Old English: *oxa*
Middle English: *stot*
Old Norse: *oxi*
**Place-names:**  Ousen Stand, Oxendale, Oxenholme, Oxen-
brow, Oxenlawe (1288), Oxenriggs, Oxenthwaite (1620), Stott
Park, Stott Ghyll

## Bull
**Derivation:**    Old English: *bula*
**Place-names:**  Bowness (Windermere), Bull Close, Bull
Copy, Bull Crag, Bullslack, Bullparrock

## Stirk
**Derivation:**    Old English: *stirc*
**Place-names:**  Strickland (1190), Strickley, Strikerigg
(1210), Stricegill (1422)

**STEER**
**Derivation:**    Old English: *steor*
**Place-names:**    Brigsteer

**BYRE**
**Derivation:**    Old English: *byre*
**Place-names:**    Byreflatts, Byers, Byrestead, Byrested (1366)

**COWSHED**
**Derivation:**    Old Norse: *fehus*
**Place-names:**    Fewsteads, Fusedale (1278), Fusethwaite, Yottenfews (1270)

# PIGS

*Bewick*

**PIGS**

**Derivation:**     Old English: *svin*

                       Old Norse: *svin, griss*

**Place-names:** Suinbanke (1180), Swindale, Swineset, Swinklebank, Swynhirst (1231), Swynthwayt (1279), Swinstonewath (1422), Grisdale, Grisebeck, Griseburn, Grisedale, Grizedale, Mungrisdale

# GOATS

*Bewick*

**Derivation:**   Old English: *gat*
Old Norse: *geit*
**Place-names:**  Gatebeck, Gatecrag (1290), Gatescarth (1210), Gatesgarth, Gatesgill, Gaterigg

### KIDS
**Derivation:**   Old English: *kidh*
Old Norse: *kide*
**Place-names:**  Kidbeck, Kid Moor, Kidsholme, Kidshowe, Kidsty Pike, Kidswell

# SHEEP

A HERDWICK RAM     *Bailey & Culley, 1790s*

**Derivation:** Old English: *sceap*
**Place-names:** Scheype Cot (1470), Shipcot, Shipman Knotts, Skiprigg

**EWE**
**Derivation:** Old English: *eowu*
**Place-names:** Ewebank, Eweclose, Ewefell, Ubank (1597), Udall (1536), Yewbank, Yewbarrow, Yewrake

**RAM**
**Derivation:** Old English: *ram*
**Place-names:** Ramsgill, Ramshaw

*The Shepherd's Guide, 1819*

## WETHER
**Derivation:** Old English: *wether*
**Place-names:** Watermillock (1215), Wetheral, Wetherbrigg, Wetheriggs

## LAMB
**Derivation:** Old English: *lamb*
**Place-names:** Lambflatt, Lambfold, Lambfoot, Lamb Howe, Lambrigg

## HOGG
**Derivation:** Old English: *hogg*
**Place-names:** Hog Hill, Hog Hole, Hoghouse, Hogg Bank, Hogs Earth

## GIMMER
**Derivation:** Old Norse: *gymbr*
**Place-names:** Gimmer Cragg

# HORSES

*Bewick*

**Derivation:**   Old English: *hors*
Old Norse: *hross, hestr* = stallion,
Old Norse: *kapall* = packhorse or small horse,
Old Norse: *jalda* = mare

**Place-names:**  Horse Beck, Horse Close, Horse Holme, Horse How, Horse Parrock; Rosber, Rosewain, Rosgill, Rosley, Rossett; Eastholme, Hestham, Hesket, Hesket Holme, Hesketh; Capplebarrow, Capple Crag, Capplethwaite, Capple Howe, Capplerigg; Hadyaud, Hodyoud, Yoadpot, Yoadscomb, Yoadcastle

# GEESE

*Bewick*

**Derivation:**    Old English: *gos* = goose
    Middle English: *stegge* = gander

**Place-names:**    Goosebutts, Goosegarth, Goosegreen, Goose-holme, Gooselands, Goosemire, Goosewell, Gosefalls, Gos-forth (1170), Goslong Sike, Stegcroft, Steglands

# 6
## BiRDS

# CRANE AND HERON

*Bewick (both illustrations) – left crane and below heron.*

**Derivation:**    Old Norse: *trani* (crane)
Middle English: *heiroun* (heron)

**Place-names:**  Trainford, Trainmoor, Trainriggs (1704), Trainthwaite (1733), Tranearth, Traneby, Tranlandes (1617), Tranmoor, Tranna(1580), Tranterne, Tranthwaite (1170), Trantrams (1241), Traynmire, Heron Island (Rydal), Heron Pike*, Heron Crag*, Herneshowe (1539)*

The crane was once a familiar bird in most parts of Britain: its presence in southern counties usually indicated by place-names derived from Old English *cran*, as in Cranbourne, Cranford and Cranbrook whereas in the north the derivation is from Old Norse *trani*. The very small number of place-names referring to the heron compared to the large number of *trani* names has led to a suggestion that perhaps in former days no distinction was made between the two but it is difficult to believe that the medieval countryman, usually so observant of natural detail, could not tell the difference between the bushy-tailed crane striding over the marshes and a crested heron standing patiently poised by a quiet river bank: he could surely distinguish as easily between these birds as he could tell 'a hawk from a handsaw'.

It seems reasonable to conclude that before the crane was hunted to extinction in England in the seventeenth century this impressive bird was a common sight in Cumbria where the numerous lakes, rivers, mosses, marshes and wetlands provided such a favourable environment.

In medieval times, despite Biblical proscription, both these birds were in great demand for the table at royal banquets and among the nobility. When in 1251 Henry III celebrated

---

*These locations may seem unlikely places to find herons and it is possible that these names may be derived from the Old English *earn*, eagle, the golden eagle or the sea eagle or the sea hawk (or osprey).

Christmas at York 115 roast cranes were served and as late as 1634 cranes appeared on the menu of a feast given in honour of Charles I.

Herons too figure quite frequently in the kitchen account books of Lord William Howard of Naworth Castle in Cumbria to which herons were delivered from heronries in various parts of the north-west. The slow flight of herons and cranes made them ideal 'targets' to train falcons and hawks and a large number must have met a sudden end in this way. In Chaucer's *Canterbury Tales* this is referred to in The Franklin's Tales where he tells of:

> *These falconers nearby a fair river*
> *Who with their hawks had many herons slain*

The invention of the 'sporting gun' and the inclination of most sportsmen and naturalists of the nineteenth century to shoot any large bird on sight resulted in the destruction of many heronries. In the mid-nineteenth century there were almost a score of major heronries in Cumbria: today there are very few, but herons may still be seen poised by almost any quiet lakeside or river bank, or at Muncaster Castle heronry or the Leighton Moss Nature Reserve.

For reasons which are far from clear the heron and the crane had by the sixteenth century acquired an aura of superstitious folklore among the lower orders of society who would neither kill nor eat them. On the contrary they became almost a protected species which it was inviting misfortune to harm, together with house martins, owls, ladybirds and daddy-longlegs.

The early years of the twenty-first century have seen a welcome increase in the heron population and the successful re-introduction to Britain of the crane, small colonies of which are now thriving on the Somerset Levels and on wetland sites in neighbouring counties as well as on the Norfolk Broads and Lakenheath Fen and also on Humberside, in Scotland and South Wales.

# CROW, JACKDAW AND ROOK

*Bewick
(three illustrations) –
top crow, right jackdaw
and below rook.*

**Derivation, Crow:**  Old English: *crawe* Old Norse: *kraka*
Old English: *hroc*   Old Norse: *hrokr*
**Place-names:**  Cracoe (1666), Cracop, Cracow, Cracalt, Crakeber, Crake Hall, Crake Trees, Crowdundle, Crowholme, Crow How, Crowmire
**Derivation, Jackdaw:** Old English: *ca*  Old Norse: *ka*
**Place-names:** Caber (1240), Kaber (1202), Kawude (13th century), Jackdaw Scar
**Derivation, Rook:** Old English: *hroc*   Old Norse: *hrokr*
**Place-names:** Rookby (1178), Rookhowe

There are probably few people today, even among those who have chosen to make their homes in the countryside, who are able to distinguish easily between a crow, a jackdaw and a rook. Yet distinct differences there are and the quite specific origins of the place-names would suggest that the farming folk who had settled in Cumbria by the early Middle Ages were well aware of them.

The grey napes and sociable habits of the jackdaws nesting in the farm buildings would not be confused with the black heads and glossy plumage of the raucous, unsociable crows nor with the white-faced, baggy-thighed colonies of rooks gathered together for their tree-top 'Parliament'. Their common partiality for a diet of grain and the young green shoots of new crops did not endear any of them to farmers of any age but most accepted that this was largely balanced by their equally voracious appetite for insects, grubs and other pests which also damaged crops.

Only the crow was positively disliked, and this was chiefly because of its preference for scavenging among any carcases it could discover and for its attacks on new-born lambs and chickens. Indeed, the Christian Church in the eighth century, with perhaps too much attention to Leviticus and too little understanding of ornithological matters, decreed that all these birds were 'unclean' and penance must be done by those who ate them. It was not until after the Reformation that the rook, at least, was cleared of this 'sin' and 'rook pie' became a common and acceptable dish, and, as we all learned in our younger days, four and twenty blackbirds baked in a pie were considered to be a dainty dish to set before the king.*

It was not the strictures of the Church but the harsh economy of the food supply which condemned rooks, crows and jackdaws alike to general and sustained hostility. An Act of Parliament of 1533 commanded every parish to equip itself with nets to catch and destroy all these birds as a threat to the grain harvests at a time of shortage and sudden increase in population. The widespread slaughter of the next two or three hundred years brought about only a temporary check in numbers and today they seem to thrive as never before.

The jackdaw, meanwhile, became, like the raven, a bird of ill-omen. The House of Commons Journal for 1604 records that a Bill under debate was immediately abandoned when a jackdaw flew into the Chamber. It was in this same year that a jackdaw was held responsible for the destruction of Holme Cultram Abbey in Cumberland, an event described in detail in William Hutchinson's *History of Cumberland* (Vol. ii page 333)

---

*An eighteenth century cookbook refers to a custom in former days of baking a pie into which live blackbirds were later placed and when the pie was opened the birds flew out singing and the beating of their wings extinguished all the candles, leading, as this account puts it, to *much diverting Hurley-Burley among the guests in the Dark.*

# CUCKOO

*Bewick*

**Derivation:**  Old Norse: *gaukr*
**Place-names:**  Gaukehau (1150), Gawk Cove, Gawklands, Gouke Park, Gowk Hill, Gowkstone

Throughout northern England, the cuckoo is known as the *gowk*, a name derived from the Old Norse *gaukr* and clearly reflected in these Cumbrian place-names. As with other migratory birds, the cuckoo's sudden arrival in the spring and its mysterious disappearance a few months later made it the subject of much superstition and curious folklore. Even Dr Samuel Johnson believed that cuckoos (and swallows) spent the winter months buried under river beds or hanging in caves.

For centuries the cuckoo was protected from harm as a

creature of an unknown world which it would be unwise to affront or injure in any way. It was a bird of mystery, often heard but rarely seen, and with a god-like propensity to create havoc in the homes of others. It was not without reason that the husband of an unfaithful wife was called a cuckold, a term well-known in Shakespeare's time – a song to the spring in *Love's Labour's Lost* includes the lines:

> *The cuckoo then, on every tree,*
> *Mocks married men; for then sings he,*
> *Cuckoo;*
> *Cuckoo, cuckoo; O, word of fear,*
> *Unpleasing to a married ear!*

It was an appropriate and ribald sense of humour which gave its name to the cuckoo-pint; it was more than rustic simplicity which found cuckoo-spit on lavender and rosemary, symbols of love and love-making. Liberal shepherds may have given grosser names to many plants, flowers and other forms of wild life than later ages found acceptable but their observations were full of meaning.

Above all, however, the eagerly awaited arrival of the cuckoo, 'the merry cuckoo, messenger of spring', signalled the end of the long, hard winter. In days when winter could mean real hunger and suffering for both man and beast, this was a profound and blessed relief. So much so that, according to one much-quoted classic tale of rustic ignorance, the folk of Borrowdale tried to build a wall round their valley to prevent the cuckoo from flying away, thus ensuring for themselves the joys of an eternal summer. The cuckoo at its appointed time skimmed over the top of the wall to frustrated cries of 'Ah! another carse would'a done it!' This improbable tale is also told of several other valleys across the north of England and was part of the repertoire of every tour guide. Many bizarre superstitions were associated with the cuckoo; e.g. cuckoo spit was poisonous; the cuckoo was really a hawk in summer plumage; if on first hearing the cuckoo a girl then counted the number of 'cuckoo notes' which followed this would be the number of years before she was married; the cicada bug is bred from cuckoo spit.

Be that as it may, the cuckoo still holds a special place in our affections as the harbinger of summer and we are still glad to welcome the voice of the first cuckoo and to share the evident joy of the unknown thirteenth century poet who rejoiced that:

> *Sumer is a-cumen in,*
> *Llude sing cuccu.*

William Wordsworth expressed the sentiments of country folk from any age when he wrote of the cuckoo:

> *O Blithe newcomer! I have heard,*
> *I hear thee and rejoice!*

# DIPPER

*Bewick*

**Derivation:**    Middle English: *dowker*
**Place-names:**  Dukerdale

This attractive little bird is usually associated with the fast-flowing streams of the uplands but in Lakeland it is most often seen bobbing up and down on the rocks in the tree-lined rivers of the valleys. For this is where the caddis-flies, the larvae and the fish eggs on which it mostly feeds are plentiful, whereas the stony beds of the higher becks are swept bare of vegetation and animal life by frequent flood.

Dippers are not present in large numbers in Cumbria but they may be found covering their territories along many of the faster flowing waters throughout the region. Regarded neither as food nor predator, as far as people were concerned, the dipper attracted little attention from the early settlers and, even in the medieval period, its haunts remained without a name. The only place-name in its honour in Cumbria is first recorded in the early eighteenth century and comes from the remote eastern fringe of the county where the dipper is still known as the 'dooker', a name directly derived from the Middle English *dowker*, related to the modern Norwegian verb *dukke*, to dip or to dive under water, an apt term to describe the dipper's remarkable ability to dive, swim and fly under water in search of its food.

Titty in Arthur Ransome's *Swallows and Amazons* must have had many a young medieval predecessor who like her lay flat on her stomachs to see the dipper bob at her and fly under the water.

# GOLDEN EAGLE AND SEA EAGLE

*Golden Eagle, left, and Sea Eagle, right, Bewick*

**Derivation:** Old French: *aigle*
Old English: *earn*
**Place-names:** Eagle Crag (Borrowdale), Eagle Crag (Easedale), Eagle Crag (Ennerdale), Eagle Crag (Grisedale), Eagle Crag (Riggindale), Erin Crag (Coniston), Erne Nest Crag (Deepdale), Heron Crag (Borrowdale), Heron Crag (Eskdale), Heron Crag (Langstrath), Heron Crag (Riggindale), Heron Pike (Glenridding), Heron Pike (Ryedale)

The return in the mid-twentieth century of the golden eagle to the Lake District and the re-introduction of the sea eagle in the

Western Isles occurred some 200 years after both these majestic birds were driven out of Cumbria by constant persecution.

From the seventeenth century onwards there is written evidence to add to the place-names themselves to prove that both these eagles nested among the crags of Buttermere, Borrowdale, Martindale and Patterdale and, also, as Thomas Pennant tells us, in many places in the mountains at the head of Windermere. The place-names suggest that they also frequented the Coniston Fells, Ennerdale and Riggindale.

The last sighting until 1957 of a golden eagle was recorded near Helvellyn in the 1830s while the sea eagle was last seen at about the same time over Black Combe, but some fifty years before that almost every author of a Lake District Guide had something to say about the eagles of Borrowdale and the famous 'Borrowdale rope' which was kept in the valley to lower intrepid young men down the face of the crags to rob the eyries of eggs and fledglings to be sold to wealthy collectors at high prices.

William Green, in his *Tourist's New Guide* (1819), asserts that the last pair fled from Borrowdale to Eskdale in the 1780s and one was shot there and the other fled and never returned. John Aubrey in 1692 and James Clarke in 1787 wrote of sea eagles on the crags near Hawes Water. According to Clarke a nest at Wallow (Walla) Crag in Borrowdale contained 35 fish and seven lambs.

It was a widely-held belief that both the golden eagle and the sea eagle carried off new-born lambs which earned them the farmers' hostility and it is chiefly for this reason that their return has not been greeted with unanimous enthusiasm. The splendid sight of these magnificent birds soaring high over the fells is one of the glories of the Lake District National Park but, in an area where lambs are now far more common than the golden eagle's usual diet of grouse, rabbits and hares, we have yet to solve the problem of reconciling the ideals of a National Park in a small and overcrowded island with the economic interests of the people who live and work there. At present it cannot be said that either the golden or the sea eagle has an auspicious presence in the Lake District.

Saxon and medieval chroniclers perpetuated and elaborated on various beliefs about the eagle first recorded in Classical Greece and Rome. The eagle's eyes were so strong that they could look directly at the sun and as old age approached the eagle rose towards the sun to burn off its ancient feathers and then plunged directly into a lake where its youth was restored. This was presented by the Christian church as proof of the sacred nature of the eagle as the only creature which could look into the eyes of God and survive as a symbol of the Resurrection.

This sacred aura was further shown in the belief that the eagle held each of its young to test its eyes against the sun and any who failed were cast out of the nest: those which succeeded were symbolic of the souls taken into heaven and those which failed were the unworthy sinners destined for Hell. Shakespeare refers to this traditional belief in an exchange between Richard (later Richard III) and Edward (later Edward IV):

> *If thou be that princely eagle's bird,*
> *Show thy descent by gazing 'gainst the sun.*

The eagle was clearly endowed with unusual powers, able to rise above earthly things, a symbol of God's message to mankind. In every Anglican church the Bible is placed on the outstretched wings of a golden eagle.

The history of the eagles is slightly linked to the story of falconry, a 'sport' which was at first the preserve of kings and their entourage of the more important nobility. Ethelbert, the Saxon King, is recorded as a devotee as was Alfred the Great, and the Bayeux Tapestry shows King Harold with a hawk. They were succeeded as enthusiastic falconers by almost every monarch until George III in the eighteenth century. The various ranks of feudal society were allocated a particular bird of prey according to their status.

It was considered a felony to own a falconry bird inappropriate to your position in society. Thus the eagle was reserved for Emperors; Kings sported the gyr falcon, Earls peregrines, yeomen goshawks, and so on down to priests who had to make do with sparrow-hawks and the poorest landowners with the lowly kestrel. One medieval account claims that at one time it was not possible to walk along the street without meeting someone with a bird of prey on his or her wrist, but it would almost certainly not be an eagle.

# GOSHAWK

*Bewick*

**Derivation:**    Old English: *gos hafoc*
                   Middle English: *gose hauke*
**Place-names:**  Goshawkstone (1723)

This splendid bird of prey is now a rare sight in Britain having
been persecuted virtually to extinction by the marauders, egg
collectors and sportsmen of the eighteenth and nineteenth centuries. It was probably never very numerous in this country
and, as an impressive killer, it was much prized in the hey-day
of falconry. It was not a high-ranking bird such as the falcon
or the peregrine and therefore not considered fit for a noble-
man or a knight but well-suited to the prosperous yeoman.

233

Two medieval documents from Cumbria shed light on the value placed on the possession of goshawks: a grant of land in the twelfth century by Lady Alice de Rumeli to the monks of Furness refers specifically to the status of the goshawks there; and a lawsuit of 1256 was concerned, among other items, with the eyry of goshawks in Thomas's Wood in Bastonwayt. Thus we have firm historical evidence to support the suggestion given by an isolated place-name that the goshawk, while not a common bird even in medieval Lakeland, neither was it entirely unfamiliar even if only as a tamed bird of prey, renowned for its handsome appearance, its speed, rapid manoeuvrability among trees and its efficiency as a killer of other birds and small mammals.

Its prey includes pigeons, partridges, grouse, waterfowl and any small mammal which may cross its path. Its name actually means 'goose-hawk' implying that at one time it was known for its ability to seize large birds like wild geese. After many years of persecution and robbing of its eggs the decline of this magnificent bird now seems to have ended and numbers are slowly increasing and there is now a significant population in Kielder Forest in Northumberland.

# KITE OR GLEAD

*Bewick*

**Derivation:** Old English: *cyta*
Old English: *gleoda*
**Place-names:** Kitcrag, Kitt Gill, Kitty Crag,* Glade How,
Glede How (Gledhill 1366)

The kite – or *glead* as it is known in northern counties – became extinct in England in the 1870s. Perhaps a dozen pairs are known to have survived in the mountains of central Wales but it is now well over a hundred years since the Rev. H. A. Macpherson claimed to have seen a kite gliding near the

---

* These names could also be derived from the personal name Kit
(Christopher)

235

Lakeland hills and probably 150 years since the last one is known to have nested there. Fortunately, recent re-introductions of the red kite to several areas of Britain are proving successful and some 2,000 breeding pairs appear to be securely established, some of them in Grizedale Forest.

Classical and medieval folklore looked upon the kite as one of the birds of ill-omen, a portent of misfortune or even disaster. In Shakespeare's *Julius Caesar* Cassius laments that the eagles have deserted the army and:

> *In their steads do ravens, crows and kites*
> *Fly o'er our heads and downward look on us,*
> *A canopy most fatal, under which*
> *Our army lies ready to give the ghost.*

Kites were also unpopular as a result of their inclination to steal small items – *snappers-up of unconsidered trifles* as the rogue Autolycus in *The Winter's Tale* describes them and warns: *When the kite builds, look to lesser linen.*

Yet in the sixteenth century kites were protected by Royal Decree for they were seen as the main scavengers of street refuse which, in those days, would comprise all manner of foul-smelling rotting waste. They became such a familiar part of the urban scene that they were even known to swoop down to take food from children's hands. This and the kite's undiscriminating diet made it disliked among the populace in general, an antipathy reflected in King Lear's outburst to Goneril: *Detested kite, thou liest.*

Royal protection disappeared in the late seventeenth century when street-cleaners arrived on the scene and sold the manure to the farmers as fertilizer. Kites were henceforth regarded, like many other wild birds, as a threat to the food supply and so were to be exterminated as 'vermin'.

In 1800, however, Dorothy Wordsworth could still write of the 'kites sailing above our heads' in Greenhead Gill, Grasmere, as if this were an everyday sight, and, a few years before, James Clarke in his *Survey of the Lake* wrote of the kite as a native of this country. Towards the end of the nineteenth century elderly folk could recall that kites once nested at Castle Head by Derwentwater, by the Ferry House on Windermere, by Ullswater and in the Eden Valley. As with so many other birds and animals kites were shot to near extinction by zealous gamekeepers and sportsmen. The return of the kite to the skies of Lakeland is a noteworthy success for modern conservation.

## LAPWING AND GOLDEN PLOVER

*Bewick*

**Derivation:**   Old English: *hleapewince**
Dialect: *tewit*
Middle English: *pluver*

---

\* The Old English word *hleapwince* meaning leap-wing refers to the lapwing's slow twisting flight when it rises from its nest.

**Place-names:** Plover Rigg, Tewit, Tewit Field, Tewit How, Tewit Tarn, Tewfitt Mires, Tewsett Pike, Tyquitemire

The lapwing or peewit (known as the *tewit* in northern counties) is one of the best loved birds with its green iridescent plumage, its plaintive cry and its spectacular tumbling flight. Until fairly recent times it could be seen almost everywhere in Britain but extensive depredations by egg collectors in the past 150 years and the widespread drainage of the marshes and moorlands which are its favourite habitats brought about a serious decline in numbers; and that decline has turned into a catastrophic collapse in recent decades with intensive use of pesticides in modern agriculture which destroy their main food supply, especially leatherjackets and insect larvae, and cause their eggs to be infertile.

Modern agricultural machinery also crushes their nests and often the young and the eggs in them. It is estimated that lapwing numbers have fallen by as much as 80% in the past half-century. To an earlier generation of farmers they were valued as voracious devourers of every type of unwanted pest in the soil.

The catalogue of destruction of birdlife revealed in parish records and in the shooting memoirs of the Victorian age makes depressing reading and it is a relief to find a gentle passage in Emily Brontë's *Wuthering Heights* where Catherine relates that Heathcliff had set a trap by a lapwing's nest containing six young, and the old ones dared not come so the young birds died. She made him promise he'd never shoot a lapwing after that, and he didn't – an interesting touch of humanity in an age when the common reaction to spotting a wild bird was to shoot it.

As with so many other birds lapwings were once considered appropriate as a source of food. The bones of lapwings have been found in the rubbish tips of Viking York (Jorvik) and on many other medieval sites. A written record also tells us that in 1634 Lord Spencer of Althorp considered a menu of wild birds, including lapwings, as an appropriate banquet to set before King Charles I. Shakespeare refers to a tradition that the lapwing chick is in such a hurry to enter the world that it runs from the nest with the egg-shell still on its head, a belief used in an amused reference to the rather callow and foolish young Osric: Horatio comments to Hamlet, *This lapwing runs away with shell on his head*, a description which most of the audience would fully appreciate.

The Cumbrian place-names refer to the northern dialect word for the lapwing – the tewit – and this attractive bird may often still be seen at Tewit Tarn in the Vale of St John and on the Pennine moors.

The single place-name referring to the plover probably refers to the golden plover which may never have been numerous in Cumbria, although Bewick tells us that in his day (1753-1828) it was abundant throughout northern Britain. It is certainly much rarer today.

# LARK

*Bewick*

**Derivation:**    Old English: *lawerce*
                     Old Norse: *læverki*
**Place-names:** Laverick, Laverickstone, Laverkewode
(1259), Laverock Bridge (1534), Lavertick, Layrocsyk (1203)

Poets of every age have taken the skylark to their hearts. From
Chaucer to Clare, from Shakespeare to Shelley and
Wordsworth, all have found inspiration in its soaring flight and
captivating song. The fact remains, however, that until early
modern times most country folk had a far from romantic view
of the lark: for them Wordsworth's ethereal minstrel of the sky
and Chaucer's messager of day, was no more than a source of
food at times when other meat was scarce or priced beyond
the means of their impoverished lives.

241

Thomas Pennant, writing in 1812, commented that poverty explained why our ancestors were as great devourers of small birds as the Italians are at present. This was probably true for there is, after all, very little meat on a small bird 'and it is scarce worth the dressing' as John Ray, the seventeenth century naturalist, put it – unless, of course, one was in really desperate need. For the wealthy, however, larks had long been a popular dish: the Lords of Naworth Castle in Cumbria were quite unacquainted with poverty, yet between 12 September and 3 October 1618 no fewer than 138 dozen (1,656) larks were consumed at their tables.

Pennant was mistaken in his implied belief that eating small birds was a thing of the past in England in 1812: a century later than this in the years before the First World War, thousands of larks were regularly on sale in the London poultry markets. Perhaps it is the unprecedented level of national prosperity in recent decades and the year-round availability of all the food we need which has helped to change attitudes to wildlife – at least in most countries in the developed world.

Sadly, this has not created favourable conditions for the lark. The RSPB reports that the skylark population in the UK has declined by more than 75% in the past 50 years, largely due to changes in farming practice. Autumn, rather than spring, sowing of cereal crops deprives the lark of its preferred nesting habitat; intensive cattle and sheep grazing inhibits growth on grassland and results in nests and eggs being made visible to predators; the use of pesticides robs the birds of a main food supply and nests can be destroyed by heavy machinery.

Despite all the casualties inflicted on the wild bird population in earlier centuries the lark remained one of the most numerous songbirds in the country and also among the most esteemed. The place-names show that, for whatever reason, the medieval inhabitants of Cumbria noted the skylark's haunts and named them. Indeed, the name was soon adopted as a personal name: one William Laveroc flourished in Westmorland in 1243, and variations on the name appear in many other counties.

Like Thomas Bewick in his *History of British Birds* they may have been impressed by the lark's keenness to be both seen and heard:

> Among the various kinds of singing birds with which this country abounds, there is none more eminently conspicuous than those of the Lark kind. Instead of retiring to woods and deep recesses, or lurking in thickets, where it may be heard without being seen, they are generally seen abroad in the fields; it is the only bird which chaunts on the wing, and while it soars beyond the reach of our sight, pours forth the most melodious strains, which may be distinctly heard at an amazing distance.

Robert Browning expressed his admiration for the lark and the universal delight in its song in words which would have been re-assuring to simple folk at any time and in any age:

> *The year's at the spring...*
> *The lark's on the wing...*
> *God's in his heaven –*
> *All's right with the world.*

Clearly a bird worthy of a place-name or a surname or, if needs must, to be baked in a pie. The lark was for many years known as the 'laverock' (from the Old English *lawerce*). Chaucer in the fourteenth century used the term when he wrote of:

> *Many flokkes of turtles and laverocks*

But soon after the mid-seventeenth century when Izaak Walton, in his *Complete Angler*, sat by a river hoping to see: 'a blackbird feed her young, Or a laverock build her nest' it was mainly used in Scotland and as a dialect word in the northern counties of England.

# MAGPIE

*Bewick*

**Derivation:**   Old French/Latin: *pi*
                Middle English: *pie, piot, pyet*
**Place-names:**   Pienest, Piot Crag, Pye Greave, Pye Howe,
Pyeing, Pyenestebanke (1295), Pyetmire, Pyet Tarn, Putholes,
Puthorns

The first element – mag – of the name 'magpie' is an abbreviation of Maggie, a diminutive of Margaret, and seems to have become attached to the medieval word *pie* or *piot* at a fairly late date. *Pie* is an abbreviation of *pied*, a reference to the bird's black and white plumage. By the early seventeenth century when Shakespeare was writing *Macbeth* the bird was generally known as the 'maggot-pie' and had already acquired its somewhat sinister reputation:

> *Augurs and understood relations have*
> *By magotpies and choughs and rooks brought forth*
> *The secret'st man of blood.*

Superstitions gathered round the bird throughout the Middle Ages and many persist to the present day. The Christian church reacted to the pagan belief that the magpie had a supernatural influence on the affairs of humanity by condemning it as the bird which sat on Christ's Cross and showed total indifference to his suffering, thus indicating that its 'magical' powers were from the Devil.

This encouraged the superstitious beliefs which have been associated with the magpie throughout the centuries. To sight a single magpie – *One for sorrow* – is considered to be a particularly bad omen and it was usual for people to cross themselves, or, later, to raise their hats, or even to spit three times, in order to circumvent the evil which must surely come their way. Magpies flying near to a window foretold death; loudly chattering magpies warned of unwelcome guests; superstition also linked magpies with the misfortunes inflicted on humanity by witchcraft. The only hope of good luck came if a magpie fled from you, as the old nursery jingle assured the young:

> *Magpie, Magpie, flutter and flee,*
> *Turn up your tail and good luck come to me*

For many years the magpie was persecuted as a pest and its numbers were seriously depleted but it seems probable that at the time when its haunts were given names this was as a result of its superstitious associations rather than from any intention to exterminate it.

Of particular interest (and suspicion) was its habit of imitating the human voice and, as Bewick put it, its addiction to stealing and hoarding bright objects. It was not until the late eighteenth century with the first stirrings of a greater understanding of wildlife and the need to protect it that the magpie's useful function of controlling the many pests which attacked the farmers' crops was once more recognised and the fear and superstition which had for so long surrounded this unusual bird began to fade.

## OSPREY OR FISH HAWK

*Bewick*

**Derivation:**     Old French: *ospriet*

There are no Cumbrian place-names directly referring to the 'osprey' as this word is not recorded before 1460\*. In the Middle Ages this bird was known as the 'Fish Hawk' or the 'Fish Eagle' and it is difficult to know which of the many 'eagle' names refer to the osprey. Names close to a body of freshwater

---

\* 'Osprey' appears in the 1611 Authorised Version of the Bible where it is named as an 'abomination' forbidden for human consumption in the dietary restrictions listed in Leviticus, a religious proscription presumably observed by many during the church-dominated Middle Ages.

containing plenty of fish are more likely to refer to the osprey.

The osprey has had a chequered history in this country. In earlier centuries it was regarded with a certain awe, chiefly because of the osprey's extraordinary skill in catching fish from the lakes and for its mysterious appearance in the early spring and its sudden disappearance in early autumn.

Inevitably it became the subject of various superstitions, including a belief that when fish saw it diving towards them they were mesmerized by some magic spell and turned belly-up in helpless surrender. We now know that the osprey's remarkable efficiency in catching its prey is due to its extraordinary eyesight and to the special alignment of its toes, two in front and two behind, the latter reversible so that slippery fish can be more firmly gripped. Shakespeare, as so often, revealed his familiarity with the folklore of the time when, in Coriolanus, he wrote:

> *I think he'll be to Rome*
> *As is the osprey to the fish, who take it*
> *By sovereignty of nature*

It was honoured in the emblems of heraldry as a white eagle. This, however, did not exempt the osprey from being included in the Act of 1566 for the statutory extermination of 'vermin' whose annoyance humanity would best be quit of, with the result that osprey numbers plummeted from indiscriminate trapping, shooting and nest destruction. . By the early twentieth century it was extinct in England. The few survivors in Scotland and Wales had to face the hazards of egg-collectors and infertility caused by agricultural chemicals.

The osprey returned to England in the 1950s by the small lake in the Rutland Water Nature Reserve but it was not until

2001 that the first breeding pair appeared in the Lake District at a nesting site by Bassenthwaite Lake. Since then the birds have returned each year, their safety and welfare being carefully monitored and great public interest being generated by the creation of special viewing points and by a video link at the Whinlatter Forest Centre direct to the nest.

The success of the Lake District Osprey Project has been dramatically demonstrated in 2015 with the arrival of new osprey chicks not only at Bassenthwaite but also at two and possibly three more nesting sites – at Foulshaw Moss in the Lyth Valle, at Esthwaite Water, and at Roudsea Moss, a few miles south of Coniston Water.

# OWL

*Bewick –*
*barn owl*

**Derivation:**    Old English: *ule*
                    Old Norse: *ugla*
**Place-names:**   Owl Gill, Owl Hirst, Uglygill, Ulbergh, Ulcat
Row (1250), Ullcoats (1205)

Of the different types of owl found in Britain it seems most likely that these Cumbrian place-names refer either to the barn owl or to the tawny owl. The little owl is a comparatively recent introduction to this country and neither the long-eared owl nor the short-eared owl has been common in northern parts at any time.

The place-names also reflect the usual habitats of the barn owl and the tawny owl, for both prefer lightly wooded sites and both choose to nest in tree holes or in buildings. Their diet of voles, mice, small birds and insects would be well provided for in the woodlands and around farm buildings.

For much of human history the owl has been regarded as an ominous and somewhat sinister bird. This may have had something to do with its nocturnal habits and its silent flight through the moonlight or with its blood-chilling shriek or its startling 'kee-vit' and ghostly 'too-whooo-ooo'. Lady Macbeth was startled in her midnight wanderings by *the owl that shriek'd, the fatal bellman*; and Geoffrey Chaucer no doubt echoed the popular superstition in fourteenth century England when he wrote of the owl as a prophet of woe and mischance.

Even in the early nineteenth century a gentleman from Devon felt impelled to inform Thomas Bewick that the 'tremulous hooting' of the tawny owl had a 'subdued and gloomy shivering… which is peculiarly horrific.' Many believed that owls could give signs of a change in the weather and that a nocturnal screech could be a warning of bad luck or even a death. When Macbeth returns from murdering Duncan he says to Lady Macbeth: *I have done the deed. Did'st thou not hear a noise?* To which Lady Macbeth replies: *I heard the owl scream and the crickets cry.*

Belief in these superstitions was quite undiminished by the efforts of the clergy to brand them as paganism and to insist that God did not use the natural world to signify his will.

It was not until the mid-twentieth century that popular belief in many of these ancient superstitions began to fade. In Victorian times owls were still to be seen nailed to barn doors to ward off lightning or evil spirits; a screeching owl was a sure weather forecast that stormy weather was on the way; a broth of owl eggs was the best medicine to give to children suffering from whooping cough; and an addiction to alcohol could be cured by a course of raw owl eggs.

By the eighteenth century the acquisition of a sporting gun had given men the means to challenge this disquieting spectre of the night and in the course of about a hundred years game-keepers and over-zealous 'sportsmen', unaware apparently of the beneficial work performed by the owl in controlling many pests, shot it into virtual extinction. Enlightened conservation is helping to restore the situation but new hazards have now appeared in the form of toxic chemicals, motorways and high-speed trains. It all seems a long way from the remote rural solitude of the thirteenth century when the owls of Ulcat Row caused the folk sleeping there to shudder in their beds.

# WOOD PIGEON OR DOVE

*Bewick –*
*wood pigeon*

**Derivation:**   Old English: *dufe*
                  Old Norse: *dufa*
**Place-names:**  Dove Cottage, Dove Crag, Dove Nest, Dovecot Wood, Dovenby, Doveslack (1201), Dow Bank, Dow Crag
(1275), Dowgill, Dowstones, Dowthwaite, Dufton

Several types of dove frequent Cumbria. The wood pigeon and
the stock dove are birds of the woodlands and valley fields,
happily making their homes in or near to farm buildings; the
rock dove makes its nest in the fissures of crags and gills and
was probably always much less numerous.

The pigeon was a regular part of the diet of many people throughout the Middle Ages as is demonstrated by the archaeological excavations of domestic refuse pits of Viking York and many medieval houses, while the surviving dovecots found on old gentry estates could house several hundred birds carefully tended for a pie or a dish of eggs. There appears to have been a tendency as the centuries passed for doves to be more restricted to the tables of the upper classes as an incident in the seventeenth century Civil War may suggest: the captain of a group of Parliamentary troops reprimanded his men for looting a gentleman's dovecot only to be met with the riposte that pigeons were fowls of the air and all men had a common right in them and they were as much theirs as the baron's. Late medieval travellers wrote of the great numbers of doves consumed in England.

The pigeon has a marked partiality for feeding on cereals, peas and beans and, as these were the main crops grown throughout the medieval period, a ravaging horde of pigeons could do immense damage and inflict great hardship on families for whom these crops could be a matter of health or hunger. Both as a source of food and as a threat to the basic food supply the pigeon was a bird of direct interest to everyone.

The differences between doves and pigeons were probably of little interest to the angry farmer or the hungry peasant but in the many references to these birds in the Bible they were quite clearly important. The dove was an elusive bird with migratory habits but the pigeon was content to remain 'at home' and so was easily kept for food.

The dove was the favoured bird partly because of its attractive plumage but mainly because it had a loving nature and appeared to be more intelligent. When the raven had failed him Noah sent out the dove to report on the flood and it returned with the information required bearing an olive leaf in its beak. This image is the traditional Christian image of peace and the dove is frequently portrayed as the human soul.

By Shakespeare's time these images of the dove were firmly established: in *Henry IV* (2) the Archbishop refers to: *The dove and the very blessed spirit of peace* and in *Romeo and Juliet* we are reminded that dove rhymes with love. It was unfortunate that these characteristics were thought to endow the dove with mystic powers and so made it a bird of superstition. However firm his religious faith or his belief in popular superstitions it is doubtful whether the medieval peasant cared greatly for the aura of sanctity bestowed on the dove when hunger stalked the land.

In medieval France the nobility and clergy were the only people allowed to keep pigeons and build dovecots. They were allowed to keep one pair of birds for every hectare of land they owned. The pigeons were kept for their eggs and for the young which were good to eat. In addition your wealth was measured in pigeons and this was sometimes used as a measurement for a suitable marriage.

# RAVEN

*Bewick*

**Derivation:**   Old English: *hræfn*
Old Norse: *hrafn*

**Place-names:**   Raven Crag (more than 30 examples), Rains-borrow Crag, Rannerdale, Ravenhals (1170) now known as Buttermere Hause, Ravens Barrow, Ravens Howe, Ravens Gill, Ravens Nest, Raven Oaks, Ravenstonedale (1223)

There is a companionable note in the raven's unmelodious 'cronk cronk' as it shadows us among the high crags, so many of which bear its name. Indeed, we have come to have a certain affection for this large and intelligent bird whose remarkable feats of aerobatics sometimes seem to be performed solely for our admiration. It was not always so.

For most of recorded history the raven has been looked upon as a bird of ill-omen, a harbinger of death, announcing by its croaking, impending calamities. Of such importance was it considered, comments the author of Bewick's *British Birds*, that the various modulations of its voice were studied with the most careful attention, and were made use of by designing men to mislead the ignorant and the credulous. The Greek classical writers handed down sufficient 'evidence' to establish the superstition that the raven boded no good to the human race, and throughout the Middle Ages it was widely held that many of the ills which afflicted the people of those times could be traced to the wicked plots of this sleek and clever bird.

Its partiality for the flesh of the corpses swinging from the many gibbets which stood so conspicuously about the countryside did nothing to enhance its reputation. Even in late Elizabethan England men believed that it was the raven which spread the dreaded plague. Christopher Marlowe, in 1592, drew a sinister picture of the awesome bird flying over town and village:

> *And in the shadow of the silent night*
> *Does shake contagion from its sable wing.*

Only a few years before this Parliament had enacted the Statute which authorised the extermination of the raven as 'vermin', a noisome and offensive creature, one of the many species of wildlife mankind would be well rid of. The justification for this judgement was the alleged losses caused to farmers by the raven's attacks on chickens, young ducks and new-born lambs.

Churchwardens were provided with funds to reward those who put the Act into effect. Wordsworth recalled seeing, when a boy, bunches of unfledged ravens suspended from the churchyard gates at Hawkshead, for which the reward of so much a head was given to the adventurous destroyer. Parish records confirm that the 'reward' was fourpence. Eventually the raven was driven to take refuge in the remote recesses of the hills where there are now probably 90 to 100 breeding pairs.

It is all a far cry from that hour of glory when Odin, the ruler of the world, chose two ravens, Hugin (thought) and Munin (memory), to be his trusted servants, to seek out information and to carry his decrees to every part of the universe. The esteem in which the raven was held in the Scandinavian world is amply demonstrated in the Norse Sagas and in the raven banners borne by the armies and retainers of King Canute, King Harald Hardrada and the Earls of Orkney.

The raven also features prominently in the mythology of many other cultures and in the English language it appears in the work of many authors from the plays of Shakespeare and Christopher Marlowe to Charles Dickens's novels, Edgar Allen Poe's well-known poem, *The Raven*, and J. R. Tolkien's *The Hobbit*. Throughout recorded history the raven has been regarded as a bird of uncanny intelligence, cunning, dexterity and foreboding.

# SPARROW

*Bewick*

**Derivation:** Old English: *spearwe*
**Place-names:** Sparrow How, Sparrowmire (1553), Sparrow Rigg

The sparrow referred to in these place-names may be either the house sparrow or the hedge sparrow or dunnock which until modern times were not distinguished, and even today there are probably few who could easily recognise one from the other. There appear to be no place-names in Cumbria referring directly to the dunnock and one only – Dunnockshaw – in Lancashire.

Yet the dunnock is common enough in the northern counties and its nesting sites are found among the juniper bushes on the fell-sides as well as in the hedgerows at lower levels. The better known house sparrow has always been well-known in cottage gardens and farmyard buildings but, even so, there are very few place-names it may claim as its own. This may well be because neither of these birds was either a primary source of food or a serious threat to crops or livestock or a danger to life and limb.

It is true that flocks of sparrows could descend on a newly-sown field and attempt to gorge themselves on the grain but in the days before compulsory school attendance there was no shortage of children available to scare them off. In Victorian times children as young as eight were employed to scare birds from the crops, spending days in the fields from dawn to dusk, alone in all weathers, for five pence a day. The children suffered more than the sparrows which fled to feed elsewhere.

Usually the sparrow prefers to feed mainly on insects of all kinds and such garden pests as caterpillars and so rarely comes into open conflict with the interests of people. It was not until the eighteenth century when the destruction of virtually any form of wildlife had become not merely a sport but in many cases a statutory obligation, that the sparrow was considered worthy of a gunshot. In 1779 a parish in Lincolnshire felt able to boast that during the year no fewer than 4,152 sparrows had been killed there.

Some sought a moral justification for this slaughter in that the Church had always frowned on the promiscuous mating habits of the sparrow, reflected in Chaucer's fourteenth century line on the Summoner: *As hot he was and as lecherous as a sparrow* and in Lucio's comment in Shakespeare's *Measure*

261

*for Measure*: *Sparrows must not build in his house-eaves, because they are lecherous.*

It was the lonely voice of the seventeenth century writer Lady Margaret Cavendish who came to the defence of the sparrow: 'What right,' she asked, 'did human beings have to shoot sparrows for taking cherries and then eat the fruit themselves?'

That few place-names refer to the sparrow suggests that in an earlier age little attention was paid to this gentle bird which never interfered with other people unless, as Charlotte Yonge commented, their colonies became numerous enough to draw attention to themselves by their noise and their squabbles, their boldness and their ubiquity.

## SPARROWHAWK

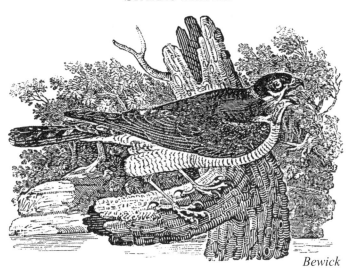

*Bewick*

**Derivation:**    Old English: *hafoc*
Old Norse: *haukr*
**Place-names:**   Hawkbarrow (1598), Hawk Crag, Hawkearth, Hawkhirst, Hawkrigg (several), Hawksdale (1285), Haucland (1280)

It is difficult to know which 'hawk' is referred to in these names. The sparrowhawk and the goshawk were mainly woodland birds frequenting Hawkhirst and Hawkdale, but Hawkrigg and Hawk Crag may refer to the falcon or the peregrine.

    Nature has few more breath-taking spectacles to offer than

the woodland pursuit of a flock of small birds by a sparrowhawk. The flash of speed and its astounding agility as it manoeuvres through the trees following every desperate twist and turn of its prey have been admired through the ages and particularly in medieval times when 'hawking' was a gentlemanly sport and when his hawk, his hound and his lady were regarded as man's only faithful friends. Chaucer reflected this aristocratic esteem for the prowess of the sparrowhawk when he wrote (with no sympathy for the tiny lark):

> *What myghte or may the sely larke seye,*
> *When that the sperhauk hath it in his foot?*

Even so, in the rigid rules of falconry the sparrowhawk ranked well below the gerfalcon, the merlin and goshawk and was generally considered appropriate for a priest. It was not until the spread of enclosed fields and the invention of the sporting gun in the late seventeenth century that hawking gradually declined. The sparrowhawk then became no longer a prized possession but a 'pest' whose destruction was authorised by Parliamentary Statute.

There followed two centuries of persecution so devastating that in 1962 Parliament had to intervene once more, this time to save the hawk from extermination by declaring it to be a protected species. Since then the principal hazard has been the widespread use of agricultural pesticides but in the woodlands and enclosed forests there has been a steady increase in numbers. The woods of South Cumbria have been especially favoured in this recovery and as the population of small mammals and birds also increases there should be food as well as safety to ensure the future of this once so highly regarded native bird of prey.

It is unlikely, however, that we shall see a return to the situation revealed in many thirteenth century landholding records from Cumbria where the rent was declared to be either an annual money payment or one sparrowhawk.

The sport of falconry was very much a gentleman's activity and the haunts of the various birds trained for the purpose were probably of greater interest to those of a higher social status than ordinary people. Shakespeare himself revealed an extensive knowledge of the technical terms of the art of falconry in his plays presumably aware that most of his audience would be familiar with the sport and so would readily understand references to haggards, towers, stoops, hoods, jesses and mews.

# SWALLOW

*Bewick*

**Derivation:**    Old English: *swalwe*
                      Old Norse: *svali*
**Place-names:**   Swalebymire (1292), Swallow Hurst, Swallow Mire, Swallow Scar, Swallow Scarth, Swallow Tail (from the shape of the field)

The swift, graceful, swooping flight of the swallow has delighted folk of every generation and its annual migration has for long been a subject of fascination. Its first sighting in the spring has always been hailed as the herald of summer while its autumnal preparations for departure inevitably bring a touch of sadness.

The disappearance of the swallow in the autumn was once a great mystery. Until two hundred years ago it was widely believed – as Dr Samuel Johnson himself believed – that during the winter months swallows buried themselves under water in

266

the beds of rivers or lakes or slept in dark corners of barns. Izaak Walton was equally convinced that they hid in caves where they hung upside down like bats and others were of the opinion that they went to the moon. Even so eminent a naturalist as Gilbert White subscribed to these ideas.

In still earlier days the swallow was the subject of much superstitious veneration, so much so that one Jacobean priest commented that 'to rob a swallow's nest is, from old beldames' catechisms, held a more fearful sacrilege than to steal a chalice out of a church.' It is not surprising that it was considered a sign of good fortune if the swallows built their nests in the eaves of one's house or barn. On the other hand augurers in Shakespeare's *Anthony and Cleopatra* are too faint-hearted to tell Antony of the doom that awaits him after the:

> *Swallows have built*
> *In Cleopatra's sails their nests.*

Admiration, fantasy and superstition may all have played a part in giving the swallow a noteworthy place in a world where Man's relationship with nature was so intimate and practical and yet surrounded with so much mystique, but which of us today, even in our materialist, hurrying, world, has not paused a moment to welcome the first swallow of the year, to wonder at its graceful flight and feel a touch of sadness when we see it flying south as summer draws to its end. Perhaps our distant ancestors had similar feelings.

The swallow was always regarded as a bird of good fortune and it was not included in the various statutes ordering the extermination of 'vermin'. John Dryden reflects this in his poetic lines:

> *Perhaps you fail'd in your foreseeing skill*
> *For swallows are unlucky birds to kill.*

267

# THE TITMOUSE

*Bewick –
great tit
above and
blue tit
below*

**Derivation:**    Old Norse: *meisingr*
                      Old English: *mase*

**Place-names:** Maize Beck, Maysingile (1190, near Asby, now lost), Maizinslack, Mazon Gill (1675), Maizon Wath

The titmouse family are among the most attractive and the most familiar of our garden and woodland birds. There can be few families, urban or rural, who are strangers to the most common of the species, the blue tit, the great tit and the coal tit, and some may even have enjoyed the sight of a flock of long-tailed tits feeding together in their garden.

The name 'tit' by which we know all these birds is derived from the Old Norse word *tittr*, used to describe any tiny object but specifically a small bird. The Old Norse and Old English words *meisingr* and *mase* (possibly originally from French *mésange*) are now quite lost to modern English and only survive in a few place-names which tell us that tits were found along the streams and in woodland glades. None of them could be considered as either a danger or as a useful addition to the food supply. No particular superstition or mythology seems to have been associated with the titmouse but it was regarded as among the most intelligent of birds.

## WHOOPER SWAN

*Bewick*

**Derivation:**    Old Norse: *elptr*
**Place-names:**    Elter Holme, Eltermere, Elterwater (1157)

In his *Guide to the Lakes* (1835) William Wordsworth refers only in general terms to the 'wild swans' on the winter lakes, and even naturalists at that time were less than certain about the differences between the three types of wild swan to be found from time to time in the Lake District.

The mute and Bewick Swans are not unknown on the lakes and tarns but they are far outnumbered by the whooper swans which may often be seen in small flocks on almost any significant stretch of water during the winter months. Elterwater in particular appears to attract regular winter visits from the whooper swans although they may also be seen on Grasmere, Rydal Water, Esthwaite Water and Windermere.

The whooper swan is larger than either the Bewick or the mute swan. It has black legs and a distinctive yellow triangle on its beak and an even more distinctive loud trumpeting call. It has powerful wings which make a swishing sound as it flies and enable it to complete the long migration route from Iceland to the British Isles in a single flight.

The swan has always been held in high esteem and it has been protected – as for example at Abbotsbury – since medieval times. An Act of Edward IV brought it under the protection of the Crown and to own swans became a mark of high status, a privilege conferred by the monarch and asserted with authority by those who held it, such as, for example, the Duke of Northumberland who forbade the inhabitants of a Yorkshire village to graze their livestock on his fenlands which were reserved for his swans.

The number of breeding swans in Britain has never been very large and it is today rather startling to read of the Victorian gentleman-naturalist who informed his readers that 'wild swans are rare in Lakeland' and on the same page reported that one day 'four of these swans passed over our heads not more than twenty feet above us' and he regretted that 'unfortunately neither of us had a gun'.

Roast swan, elaborately dressed, had appeared as a festive dish on the tables of Kings and privileged nobility throughout almost all recorded history, and in 1661 even an ordinary knight such as Sir Daniel Fleming of Rydal had to pay one shilling for a swan to adorn his Christmas feast. Sir Daniel would not have gone hungry without his swan but it enhanced his status in the society of his day. Fortunately, swans were not a common item on the menu of even the upper classes so the honking of the whoopers on Elterwater may be heard as clearly in the twenty-first century as it was in the twelfth.

It is, however, most improbable that the visitor to Elterwater will ever hear the singing of the dying swan, a legend which has persisted for more than 2,000 years and has given our language the phrase 'swan-song'. Scientific studies have proved this to be no more than a romantic fantasy but popular and poetic fancy have ensured its survival. Chaucer in the fourteenth century, Shakespeare in the sixteenth century followed by many other poets and musicians since, have kept the belief firmly enshrined in our folklore. At a decisive point in *The Merchant of Venice*, Portia chooses her words with solemn care:

> *Let music sound while he doth make his choice;*
> *Then, if he lose, he makes a swan-like end,*
> *Fading in music*

The most enduring legend of the swan is perhaps the tale, first recorded in the twelfth century, of The Swan Knight which tells of a questing knight drawn across the lake in a boat pulled along by a swan as he seeks to rescue a damsel in distress, a story told in many versions over the years and best known in modern times through Richard Wagner's opera *Lohengrin.*

Had Samuel Taylor Coleridge heard this he may not have written his barbed lines:

> *Swans sing before they die– 't were no bad thing*
> *Should certain persons die before they sing.*

# WOODCOCK AND BLACKCOCK

*Bewick (both illustrations), top woodcock and below blackcock.*

**Derivation:**    Old English: *cocc*

                     Old Norse: *orri*

**Place-names:**  Cock Cove, Cosca Hag, Cock Hill, Cock How, Cock Rigg, Cockshot (several), Cockstone (1390), Cockup, Over Water (Orre Water 1687).

The following names refer to the *lek*, the open space where the blackcock perform their mating displays: Cocklake, Cocklate, Cocklaw Fell, Cocklaye(1189), Cocklee, Cocklethwaite, Cocklet Rigg, Cockley, Cockley Beck, Cockley Gill, Cockley Moor, Cockley Moss

The Old English word *cocc* from which almost all these place-names are derived could refer to either the blackcock (the black grouse) or the woodcock. Both birds were once common in the marshes, woodlands and open moorlands of Cumberland and Westmorland, providing an impressive display with their mating rituals and their territorial flights, the woodcock noisy and theatrical in their dawn 'leking', the blackcock slow and deliberate in their twilight 'roding'.*

These performances have long attracted the attention and curiosity of man but it was for food on the table that these birds were caught in the snares set to trap them (Shakespeare's *springes to catch woodcocks*) and, eventually, in the eighteenth and nineteenth centuries, they were almost destined to extinction. For in those years a wealthy and rapidly expanding London market took vast numbers of blackcock and woodcock from the Lake District, as a passage from Thomas Pennant's 1770's *Tour of Scotland* tells us. He describes how in the hills

---

* The origin of the word roding (the evening territorial flight) is obscure. The lek (the dawn mating display) is derived from the Old Norse *leikr* meaning to play)

near Windermere there were 'numbers of springes for wood-cocks, laid between tufts of heather, with avenues of small stones on each side to direct these foolish birds into the snares, for they will not hop over the pebbles. Multitudes are taken in this manner... and sent to the all-devouring capital by the Kendal stage.'

The woodcock was, it seems, considered to be a rather stu-pid bird and so easy to entrap. It had, apparently, had this rep-utation for many years when Shakespeare's Grumio in *The Taming of the Shrew* remarked so disrespectfully:

*O! this woodcock! What an ass it is!*

In the eighteenth century formal shooting parties became a fashionable form of aristocratic entertainment and grouse, pheasants and partridges were bred and carefully protected for this purpose. By the 1770s the shooting seasons had acquired their modern forms. These sporting massacres made further inroads and with the later destruction of the heather moors and woodlands it seemed as if the black grouse and the woodcock would soon disappear entirely from the Cumbrian scene.

The movement towards conservation of woodlands and wild places and the wildlife within them might have saved the day and, although the number of places where the drama of the 'lek' or the 'roding' might now be seen is far fewer than in the list from earlier days, it is gratifying to know that the flash of red wattles and the fanning of white tails may still be seen in the dawn light or at twilight in a few of the forest glades and open moorlands of Cumbria.

# 7
## FISH

# Eel

*Hill*

**Derivation:**   Old English: *ael*
Old Norse: *all*
Early Modern English: *pod-net*
**Place-names:**   Eel Acres (1588), Eel Ark, Eel Beck, Eel Mire, Eel Tarn, Podnet, Podnett

The rivers of England teemed with eels until historically very recent times. Archaeologists excavating the waste pits of Saxon, Viking and Medieval settlements have established that both the river eel and the conger eel formed an important part of the diet of people of all classes. Written records from both before and after the Norman Conquest refer to consignments of eels, often in very large numbers, as items in the structure of feudal rents.

A document from the reign of Edward the Confessor tells of a tenant who had to produce 'foure thousand eols in Lenton to carte to the abbot' and the Isle of Ely (so-callcd, as Bede explained, because of the large number of eels which were found there) was committed, appropriately enough, to pay an annual levy of 100,000 eels due to the Crown. Ely's long relationship with the eel is continued today with the city's unique 'Eel Trail' discovering its many historic, artistic and architectural features. The cathedral is the only building in the United Kingdom to be listed as one of the 'Seven Wonders of the Middle Ages'.

In the reign of Edward I a Cumberland landowner received a request for quantities of large eels to be sent to the King's Court, at that time in Scotland during Lent. Many later writers confirm that the lakes and rivers of Cumbria abounded with eels weighing up to five pounds so it is likely that the royal table was duly provided for. This liberal consumption of eels seems not to have been diminished by the Biblical ban on its consumption, confirmed in English with the appearance in 1611 of the Authorised Version of the Bible where it is clearly stated: 'Whatsoever hath no fins or scales in the waters, that shall be an abomination unto you.' This religious prohibition was given support by the popular belief that eels fed on mud, a superstition apparently well-known to William Shakespeare and Doll Tearsheet when she berates Falstaff:

*Hang yourself, you muddy conger, hang yourself!*

In the mid-seventeenth century Izaak Walton devoted a whole chapter in his *Complete Angler* to the eel which he praised as: 'a most dainty fish regarded by some the queen of palate-pleasure.' He did, however, admit that: 'physicians account the eel dangerous meat' and quoted an Italian adage: 'Give eels and no wine to one's enemies.' Such strictures appear not to have diminished the appetite for the eel among generations of Romans, Anglo-Saxons, Norsemen, Normans and medieval Englishmen – and the modern London cockney.

The place-names suggest that many of the smaller water-courses were also well stocked with eels for those who could trap or spear them. 'Podnets', on the other hand, were used in the lakes and larger rivers, and the method of trapping eels in this way appears, like the word itself, to have been peculiar to the Lake District. It is described in an article in *Longman's Magazine* in 1892:

> In this long wall of net are three or four openings to which purse-nets, about eighteen feet long, stretched on hoops and bow-nets, are attached, the far end being closed. These 'pods' as they are called, are extended downstream and attached to stakes in the river bottom, their positions being marked by float.

Smoked eel, boiled eel, stewed eel, jellied eel and eel pie were for many centuries the food of rich and poor alike, nutritious, easy to digest and delicious with herbs. Some cookbooks still include them but we are more likely to meet the eel today as a tiny strip hiding in a piece of lettuce presented as an hors d'oeuvre. This may perhaps be attributed to increased concern about the toxicity of eels (eel blood is highly toxic to humans) but more probably to a massive decline (some say 95%) in the number of eels migrating to British waters in recent years.

Lent, as we have noted, was the favourite season for eels in earlier and more spiritually devout ages and, by a happy co-incidence, the eel could also provide another need much in demand at that time of the year. A salted eel-skin made an admirable instrument of chastisement or flagellation and, long after the monks had departed, Samuel Pepys in the late seventeenth century still found it useful for this purpose: 'With my salt eele (I) went down in the parler, and there got my boy and did beat him.'

# SALMON

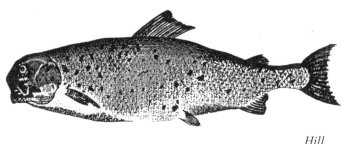

*Hill*

**Derivation:**  Old English: *leax*
Old Norse: *lax*
**Place-names:**  Lacksloft (twelfth century), Lasgill Beck, The Loups, Laxthwait (1220), Crowder's Leaps, The Leaps

The salmon was sacred to the Druids as the embodiment of wisdom and this is also reflected in the role of the salmon in Celtic, Anglo-Saxon and Norse mythology and folklore. Following the Norman Conquest of England in the eleventh century the new landowners sought to surround the salmon with an aura of legal privilege much of which has lasted until the present day.

The records of the religious houses of Cumbria, – Furness, Fountains, Calder, Holme Cultram, Lanercost, St. Bees, Shap, and Carlisle – echo to the sound of ecclesiastical lawyers battling to defend or to secure the rights of salmon fishery within their estates. The history of a thousand years is littered with bitter and costly conflicts over the right to fish this one highly prized and highly priced fish.

In Daniel Defoe's account of his travels through England in the early eighteenth century we read that salmon from the waters of Cumbria were sent to the London market: 'upon horses which, changing often, go Night and Day without intermission', to be sold there 'from 2s 6d to 4s per pound', or roughly £20-£30 in modern values. It is not at all difficult to understand why landowners were so anxious to protect their rights or why poaching became so widespread and profitable.

Mr Crowder of Crowder's Leaps was no doubt urgently engaged in such matters but most of the ordinary folk of pre-Conquest Cumbria would probably have no such concerns – for a time, at least, they could take all they needed and enjoy it, even if that involved some nocturnal poaching. The rivers of Cumbria were then full of salmon which appeared regularly on the tables of those who lived near them. The household accounts of Millom Castle in the reign of Queen Anne show that salmon featured regularly on the table of the chatelaine, Madam Bridget Hudleston.

It would seem that salmon were so abundant in the River Duddon that they were provided almost daily as a cheap food for the young apprentices at the Ulpha bobbin-mill and, in a new contract drawn up on their behalf, it was stipulated that they should not be given salmon on more than three days a week. Perhaps they had been presented too often with the rather unappetizing salmon dried in lofts such as that at Lacksloft, not a recipe Izaak Walton would have approved of: for him the salmon was the king of the fresh-waters.

# TROUT

*Isaak Walton*

**Derivation:**    Old English: *truth*
                   Old Norse: *rauthi*
**Place-names:**   Troutal, Troutbeck (several), Rothay, Routhmere (Rydal Water)

In 1892 the Cumbrian naturalist H. A. Macpherson wrote:
'The becks and lakes of this mountainous region abound in
trout, varying in size and appearance according to locality, but
everywhere affording sport in the loveliest and most romantic
spots that can be found.' These words might equally have been
written five hundred or a thousand years earlier, and even
today, the *salmo trutta* is still to be found in almost every lake,
tarn, river and beck in Cumbria.

A modern fly fisherman may be lucky enough to catch a
trout of up to three pounds; an angler using a spinner might
double that weight. It is probable that their medieval predecessors, more concerned perhaps with obtaining a good meal

than with a day's sport, caught them in dozens or in scores using a device known as an 'otter', still in use in many countries despite many failed attempts, even by United Nations resolutions, to declare it illegal.

The Norse settlers referred to the trout as *rauthi*, meaning 'the red one', and from this word are derived the names of the River Rothay and the lake which we know as Rydal Water but which until modern times was known as Routhmere, 'the lake of the red one'. It might appear obvious that this description refers to the red char but for the strange fact that the char moving through the waters of Windermere enters only the River Brathay and apparently ignores the River Rothay. Whether this has always been so it is impossible to say but Sir Daniel Fleming of Rydal Hall certainly observed in 1671: 'that he scarce ever got any trouts in Brathy or case (char) in Routha-meer.'

Long before angling for trout became the sophisticated 'art' which Izaak Walton described, medieval folk had employed a quite different technique to secure a trout for supper. This was the art of tickling which Shakespeare was clearly familiar with for he makes several references to it, as for example when in *Twelfth Night* Maria exhorts Sir Toby and Sir Andrew to hide as Malvolio approaches:

*for here comes the trout that must be caught wih tickling.*

# 8
## INSECTS

# ANT

**Derivation:** Old Norse: *maurr*
**Place-names:** Marthwaite (lost), Maureberghe (twelfth century)

The ant has always been such a common insect that only those places where its nests or 'ant-hills' were especially large or numerous and the ants were likely to invade the houses would attract particular attention. Otherwise there is little to suggest that the ant was of any great interest to ordinary folk of the Middle Ages except as a pest. However, during the political debates of the seventeenth and eighteenth centuries, the ant nests were studied and praised as a model form of government where the entire community worked together for the common good. One might surmise that the medieval ants-nests of Marthwaite and Maureberghe acquired their place-name identity only because of their size.

Unsurprisingly, there are few extant place-names referring to the northern dialect word for the ant, the *pismire**, but Pismire Lane is found in Cumbria, Pismire Beck in Yorkshire, Pismire Hill in Derbyshire, and similar names also exist in Ireland and Scotland. Pismire was well-known to both Chaucer and Shakespeare. In *The Summoner's Tale* Chaucer's 'god wyf' complains that her husband is as angry as a 'pissemyre, Though he have al that he kan desire'; and two hundred years

---

* Pismire is most probably derived from Middle English *pisse* (i.e. urine, referring to the smell of formic acid associated with anthills), and Old Norse *maurr*, an ant).

later Shakespeare's Hotspur declares in Part I of *Henry IV*:

> *Why, look you, I am whipped and scourged with rods,*
> *Nettled and stung with pismires, when I hear*
> *Of that vile politician Bolingbroke.*

# BEE

*Thomas Muffett*

**Derivation:** Old English: *beo*
**Place-names:** Beegarth, Beehive Beck

The bee was a sacred creature in Ancient Greece and Rome and this strong sense of communion with the bee continued in the culture of the Celts, the Picts, the Saxons and all the Nordic people. Bees were regarded as part of the community and would take offence if they were not treated with respect. They might refuse to produce honey or the entire hive might die.

Thus, for example, when someone in the family died it was important to tell the bees and to drape a black ribbon on the hive. They should also be given a share of the funeral repast. All family news should be discussed with them and their hives should be protected at all times.

Honey was an important item in the life of those times. It was known to have positive health benefits and to be one of the most effective healing agents and antiseptics in the treatment of wounds and burns and the gastric problems which were so common in medieval families.

It was, above all, the chief ingredient of mead, the drink known throughout Northern Europe for untold centuries. Up to about four pounds (2 kgs) of honey were required for each gallon (5 litres) of water, to be boiled then fermented with yeast, with herbs added to give a variety of special flavours.

This was the drink for special feasts and celebrations and considerable quantities of honey would be needed to provide all the mead consumed on these occasions. The literature of those times gives us a glimpse of these heroic feasts – as in the Sagas and in poems such as Beowulf where we are told that in the great mead-hall an attendant stood by '…pouring helpings of mead and famous men fell to with relish as round upon round of mead was passed.' For daily use, in more humble households, honey was widely used as a sweetener.

Eliza Linton, writing in the 1860s, relates that: 'Every summer and autumn hundreds of hives are brought up to Ennerdale and set on Revlin, for the bees to get strength and sustenance before winter time.' One such portable hive and a swarm of bees were found during the excavations at the Viking settlement at Jorvik. Bee-keeping was considered to be so vital to the welfare of a community that the specialist office of bee-keeper (Old English *bicere*) is occasionally referred to, as, for example, in place-names such as Bickershaw and Bickerstaffe in Lancashire.

The hives were made of wicker skeps and it was usually necessary to destroy these before the honey could be collected. One of our Cumbrian names 'Beegarth' suggests that this was a place where hives were kept in an enclosure, perhaps an orchard.

The skeps, which were made of straw, needed to be protected from the wind and rain and were usually housed in bee boles, alcoves or recesses constructed in stone walls. Many of these bee boles have survived in Cumbria and examples may be seen at Brantwood, Fell Foot, Hill Top, Holme Garth and Stang End. So important were the bees that it was not unusual for rents to be paid in beeswax or honey.

*A bee bole.*

## LOUSE

**Derivation:**    Old Norse: *lus*
**Place-names:**  Lousegill

The louse was too common a pest, too intimately and universally known to warrant a place-name, and it is possible that the eighteenth century place-name, Lousegill, was one of the facetious and derogatory names which it was then the fashion to impose on certain places at that time. It was not necessarily infested with lice but was just a useless piece of land. But there can be little doubt that head lice and body lice were a serious pest in medieval times when one remedy appears to have been an ointment made up largely of pork fat and incense.

By the time of Elizabeth I the problem of head lice was quite unresolved and a different solution was sought by the wealthy classes. The head was shaved and the hair replaced by a wig, a fashion adopted by the Queen herself and soon followed by most of those who could afford to do so. However this proved not to be a total answer to the problem of lice as Samuel Pepys revealed when he records in his *Diary* in March 1663 that when his wig-maker brought him his wig he found it was 'full of nits and I did send it to him to make it clean'.

Lice in wigs remained a problem but the fashion remained until in 1795 the government put a tax on wig powder and only bishops and barristers clung on to their wigs. But head lice remained as persistent a problem as it had been over 1,000 years before. A nineteenth century treatise on 'the Power, Wisdom and Goodness of God' maintained that the louse was created as an indispensable incentive to humanity to observe habits of cleanliness.

# SPIDER*

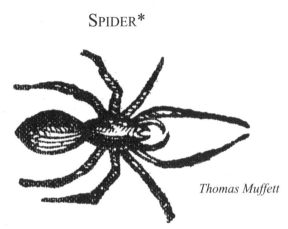

*Thomas Muffett*

**Derivation:**    Old English: *spinnan*
**Place-names:**  None known in Cumbria

There are few references in the early literature of European folklore and mythology to the spider except for the tale of Arachne, the maiden of ancient Greek mythology who was turned into a spider by her jealous rival in the art of weaving, the goddess Athena.

The spider does not figure largely in Norse mythology nor does Celtic or Anglo-Saxon literature reveal that the spider had any special significance in the mythology of those people. Yet the common name of the spider in the dialect of the northern

---

* Spiders are not insects but belong to the family arachnid. There arc many differences between spiders and insects; for example, insects have six legs and three distinct body parts while spiders have eight legs and only two body sections.

counties of England – the *attercop* – is derived directly from the Old English word *atercopper* and the Old Norse *eitr-cop,* meaning 'poison head', reflecting the belief that the spider's bite was venomous and often fatal to humans.

In *A Midsummer Night's Dream* the spider was forbidden to approach the Fairy Queen:

> *Weaving spiders come not here;*
> *Hence you long-legged spinners, hence!*
> *Come not near our fairy queen.*

Fear of the spider's supposed venom was almost universal and was strongly reinforced by the nineteenth century vogue for 'horror' which was the theme of various novels and most famously in the nursery rhyme of Little Miss Muffet's flight from a spider which sat down beside her. Over a hundred years earlier spiders were described as 'so loathsome that ladies squealed at the sight of them'.

In contrast to this centuries-old superstitious fear of the spider there is also plentiful evidence that the spider was looked upon as one of nature's special creatures and beneficial to humanity. Athena turned Arachne into a spider not in vengeance but so that she could continue her wonderful weaving for ever. The spider thus became the weaver of humanity's destiny, endowed with an aura of sanctity which was continued in the various European cultures of many succeeding centuries.

It was considered certain to bring bad luck to kill or even harm a spider and to bring good luck if a spider crossed your foot or was found in your clothes or bed linen. Spiders, unlike insects, were tolerated in the house not only as potential bringers of good fortune but as creatures with practical uses. They must have been especially welcome as a means of reducing, if only by a little, the swarms of flies which were a constant pest in every household until modern times.

Their cobwebs were widely used to stem bleeding from wounds; a spider swallowed live with butter was believed to cure jaundice; cows suffering from internal bleeding could also be treated in this way; a fever could be cured by wearing a spider enclosed in a nutshell round the neck. By the seventeenth and eighteenth centuries, at a time when Parliamentary Statute had authorised the mass slaughter of most wildlife as 'vermin', more scientific studies were beginning to create a deeper appreciation of the natural world. The spider came to wear its halo of sanctity once more. In 1653 the physician, John Bulwer, asked whether 'it were lawful to destroy any one of God's creatures though it were but… toads and spiders, because this were taking away one link of God's chain.'

Today the spider is more likely to be associated with the legendary tale of Robert Bruce or with the nursery finger-play rhyme of unknown origin:

*Itsy Bitsy spider*
*Climbing up the spout…*

than with its remarkable and beautiful weaving or the strange superstitions surrounding it; but to the children of Anglo-Saxon and Old Norse England, spiders would have been much more like household pets, and far more homely than Aragog and Shelob, the monstrous giant spiders of J. K. Rowling's *Harry Potter* and J. R. Tolkien's *Lord of the Rings.*

# 9
## REPTILES AND AMPHIBIANS

# HAGWORM

*Topsell*

**Derivation:**   Old English: *hagwyrm*
              Old Norse: *hoggormr*
              Old English: *snaca*
**Place-names:**   Hagworm Close, Hagworm Gill, Wormgill,
Wormhowes, Wormpots, Wormrigg, Snarker Moss, Snarker
Pike, Snake Pike

The terms 'hagworm' or 'worm' were once popularly used to
denote almost any type of snake but in Old Norse, Old English
and Modern Norwegian it refers more specifically to the viper
or adder. It is not known whether there have ever been significant numbers of these reptiles in the north-west counties of
England but William Hutchinson in his *History of Cumberland*
(1794) referred to the slow-worm, asp and hagworm or snake
and a more recent study (G. A. K. Hervey and J. A. G. Barnes
1970) reports 'the presence in Cumbria of the Adder or Viper,
the Grass Snake, the Slow-Worm and the Viviparous Lizard.'

Many Viking-age stone sculptures in the north, such as those at Dacre, Gosforth, Lowther and Penrith, show that the 'snake' was a familiar creature at that time, and either as the treacherous world-serpent of Norse mythology or as the sinister symbol of sin in the Christian religion, it had a well-established reputation as the enemy of human happiness. The most common vision of Hell in Anglo-Saxon times portrayed a tormenting pit of snakes and in Norse folktales a common punishment for those who offended those in authority was to be flung into a snake-pit: 'Naked men strive amidst the serpents… here is the hiss of adders and here serpents have their dwelling.'

Whatever form of 'snake' the hagworm may have been it was certainly not regarded as a friend of humanity. Fear of the venomous bite of the snake was, probably rightly in those days, a constant concern and was naturally connected to the fear of witchcraft which featured so prominently in the lives of country folk in the medieval centuries. The adder's fork and the blind-worm's sting were right and proper items to be cast into the witches' cauldron to make the hell-broth boil and bubble.

# TOAD

*Topsell*

**Derivation:** Old Norse: *padda*
**Place-names:** Paddle Beck, Paddockwray (1570), Paddigill (1543), Padmire, Pavement End

To many country folk the common toad is still a 'paddock', a name derived from the Old Norse *padda* and the origin of our place-name examples from Cumbria.

Appearances have weighed heavily against the paddock and, for the most part, it has been a creature to inspire more aversion than affection. Perhaps only the appreciative gardener is its friend. As early as the fifth century, Clovis, architect of the Kingdom of France, considered it prudent to change his insignia from three toads to three fleur-de-lis; by Shakespeare's day superstition had firmly established a belief that this ugly, venomous reptile was a witch's familiar (as in Macbeth) able to work magic spells, suck milk from the udders of cows and even poison them.

Most persistent of all was the belief that the toad had a jewel in its head which was the source of its magic power, a superstition which was the inspiration for Hans Anderson's tale *The Toad* and was clearly well-known to Shakespeare when he wrote in *As You Like It*:

> *Sweet are the uses of adversity*
> *Which, like the toad, ugly and venomous,*
> *Wears yet a precious jewel in his head.*

All in all the toad was regarded as a sinister creature probably a work of the Devil, and so hateful to all good Christian folk that St. Patrick had it banished from Ireland. Even the more enlightened commentaries on the natural world put forward in the eighteenth century were under strain where the toad was concerned: did Man have the right to destroy any of God's creatures – even toads? Nature, it was explained, was a complex organisation in which even each hated toad had a part; therefore it was wrong to cause harm to any living creature – even toads.

Not even the joie de vivre of A. A. Milne's *Toad of Toad Hall* has succeeded in putting the poor paddock in the same class as Peter Rabbit or Squirrel Nutkin and, regrettably, the pastime of 'spang-hewing' (catapulting) toads high into the air to fall to a messy death is still not unknown. Perhaps the most charming words ever written about the toad appear in Robert Herrick's *Grace for a Child*:

> *Here a little child I stand.*
> *Heaving up my either hand;*
> *Cold as paddocks though they be,*
> *Here I lift them up to Thee,*
> *For a benison to fall*
> *On our meat and on us all*

# BIBLIOGRAPHY

The number of books and specialist articles on the history of the Anglo-Saxon and Viking Age and on the history of medicine and on the many aspects of natural history and folklore has increased steadily in recent years. The following list represents a selection of the books which the author found most useful in the preparation of this study. Specialist articles consulted are not listed.

PLACE-NAMES:
*The Place-names of Cumberland*: Volumes XX, XXI, XXII in the publications of the English Place-name Society (1950-52)
*The Place-names of Westmorland*: Volumes XLII and XLIII (1967)
*Bede, Ecclesiastical History of the English People* (OUP 1969)
A. Bradshaw, *Ancient and Interesting Trees of Cumbria* (2003)
D. Brearley, *The Place-names of the Lake District explained for the general reader* (1978)
E. Ekwall, *The Oxford Dictionary of English Place-names* (1967)
R. Gambles, *Lake District Place-names* (2013)
J. Lee, *The Place-names of Cumbria* (1998)
A. D. Mills, *The Oxford Dictionary of English Place-names* (1998)
D. Mills, *The Place-names of Lancashire* (1976)
A. H. Smith, *English Place-name Elements* (1856)
D. Whaley, *A Dictionary of Lake District Place-names* (2006)

GENERAL:
E. A. Armstrong, *The Folklore of British Birds* (1958)
R. N. Bailey, *Viking Age Sculpture* (1980)
T. Bewick, *The History of British Birds* (1804)
T. Bewick, *A General History of Quadrupeds* (1807)
W. Bonser, *The Medical Background of Anglo-Saxon England – a*

*study in history, psychology and folklore* (1963)

C. M. Bouch, *Prelates and People of the Lake Counties* (1948)

C. M. Bouch and G. P. Jones, *The Lake Counties 1500-1830* (1962)

S. Brink (ed) and N. Price, *The Viking World* (2012)

G. Chaucer, *The Canterbury Tales* (Everyman edition 1966)

J. Clarke, *A Survey of the Lakes...* (1789)

N. Culpeper, *Complete Herbal and English Physician* (1653)

S. Denyer, *Traditional Buildings and Life in the Lake District* (1991)

D. Fleming, *Description of Cumberland, Westmorland and Furness* (1671)

P. G. Foote and D. M. Wilson, *The Viking Achievement* (1980)

John Gerard, *The Grete Herbal* (1633)

W. Gilpin, *Observations...* (1786)

M. Grieve, *A Modern Herbal* (1931)

J. Guest and M. Hutcheson, *Where to Watch Birds in Cumbria* (1997)

G. Halliday, *A Flora of Cumbria* (1997)

R. Hall, *The Excavations at York: The Viking Dig* (1984)

J. E. Hartung, *British Animals extinct within Historic Times* (1880)

S. Heaney, *Beowulf: A new translation* (1999)

G. A. K. Hervey/J. A. G. Barnes, *Natural History of the Lake District* (1970)

J. Hill, *An History of Animals* (1752)

W. Hutchinson, *The History of Cumberland* (1794)

Gwyn Jones, *A History of the Vikings* (1968)

J. de Bairacli Levy, *The Illustrated Herbal Handbook* (1978)

T. Machell, *Antiquary on Horseback* (ed. J. M. Ewbank, 1963)

H. A. MacPherson, *A Vertebrate Fauna of Lakeland* (1892)

T. Moffet, *The Theater of Insects* (1658)

J. Nicolson and R. Burn, *The History and Antiquities of the Counties of Westmorland and Cumberland* (1777)

N. Oliver, *Vikings, A History* (2012)

W. H. Pearsall, *Place-names as clues in the pursuit of Ecological History* (1961)

W. H. Pearsall and W. Pennington, *A Landscape History* (1973)

W. Pearson, *Letters, Journals and Notes on the Natural History of Lyth* (1844)

R. C. A. Prior, *On the Popular Names of British Plants* (1879)

D. Ratcliffe, *Lakeland, The Wildlife of Cumbria* (2002)

W. Rollinson, *Life and Tradition in the Lake District* (1974)

W. Rollinson, *The Cumbrian Dictionary of Dialect, Tradition and Folklore* (1997)

S. Rubin, *Medieval English Medicine* (1974)

Snorri Sturluson, *Edda*: translated A. Faulkes (1995)

C. Swainson, *The Folklore and Provincial Names of British Birds* (1886)

C. H. Talbot, *Medicine in Medieval England* (1967)

K. V. Thomas, *Man and the Natural World* (1983)

E. Topsell, *A History of Foure Footed Beastes* (1607)

William Turner, *Herbal* (1558)

P. Walker and E. Crane, *Bee Shelters and Bee Boles in Cumbria* Transactions CWAAS 1991).

Izaak Walton, *The Complete Angler* 1653 (1945 edition)

Thomas West, *A Guide to the Lakes* (1778)

Gilbert White, *Natural History and Antiquities of Selborne* (1789)

Dorothy Wordsworth, *Journals* (OUP edition 1978)

William Wordsworth, *Guide to the Lakes* (1835: OUP edition 1970)

J. Wright (ed), *English Dialect Dictionary* (1898-1905: reprint 1970)

FIELD GUIDES:

The Reader's Digest, *Nature Lover's Library*

Oxford University Press, *Oxford Books of Birds, Wild Flowers, Trees*

Frederick Warne, *Observer Books*

Collins: *Birds of Britain and Europe*
       *Mammals of Britain and Europe*
       *Trees of Britain and Northern Europe*
       *Wild flowers of Britain and Northern Europe*

# ABOUT THE AUTHOR

ROBERT Gambles was born and grew up in Derbyshire. He was a scholar of St John's College, Oxford, where he took an Honours degree in Modern History and a post-graduate Diploma in Education. He also has a Licentiate Diploma in Music. His professional career was spent in Education, mainly in Ely and Liverpool.

He acquired a love of the Lake District early in life and he has lived in Cumbria in his years of retirement during which he has explored the whole district and written a number of books and many articles on various aspects of its history.

The author has also pursued his interest in a wider national history and a critical study of some of the well-known stories from British history was published in 2013 under the title Great Tales from British History, and was decribed by The Guardian as 'hugely enjoyable'. Through his Norwegian wife he acquired a special interest in the life and history of Norway: Hayloft recently published his acclaimed Espen Ash Lad a collection of Folk Tales from Norway.

A keen but pragmatic interest in conservation and the protection of the natural environment has always featured in his philosophy of life and he was for many years a Trustee and member of the Executive Committee of the Friends of the Lake District. He has also worked as a volunteer for the National Trust.

# BOOKS BY THE SAME AUTHOR

*Man in Lakeland* (1975)
*Exploring the Lakeland Fringe* (1989)
*The Spa Resorts and Mineral Springs of Cumbria* (1993)
*Walks on the Borders of Lakeland* (1995)
*Yorkshire Dales Place-names* (1995)
*Walks Round Windermere* (1997)
*Echoes of Old Lakeland* (2010)
*Escape to the Lakes: The First Tourists* (2013)
*Great Tales from British History* (2013)
*Lake District Place-names* (2013)
*Espen Ash Lad: Folk Tales from Norway* (2014) translation
*The Lakeland Dales* (2016)